Holt Geometry

Homework and Practice Workbook

Teacher's Guide

HOLT, RINEHART AND WINSTON
A Harcourt Education Company
Orlando • Austin • New York • San Diego • London

Copyright © by Holt, Rinehart and Winston.

All rights reserved. No part of this publication may be reproduced or transmitted in any form or by any means, electronic or mechanical, including photocopy, recording, or any information storage and retrieval system, without permission in writing from the publisher.

Teachers using GEOMETRY may photocopy complete pages in sufficient quantities for classroom use only and not for resale.

HOLT and the **"Owl Design"** are trademarks licensed to Holt, Rinehart and Winston, registered in the United States of America and/or other jurisdictions.

Printed in the United States of America

If you have received these materials as examination copies free of charge, Holt, Rinehart and Winston retains title to the materials and they may not be resold. Resale of examination copies is strictly prohibited and is illegal.

Possession of this publication in print format does not entitle users to convert this publication, or any portion of it, into electronic format.

ISBN 0-03-078088-8

4 5 6 7 8 9   862   09 08 07

Contents

Blackline Masters

1-1 Practice B	1	6-6 Practice B	42
1-2 Practice B	2	7-1 Practice B	43
1-3 Practice B	3	7-2 Practice B	44
1-4 Practice B	4	7-3 Practice B	45
1-5 Practice B	5	7-4 Practice B	46
1-6 Practice B	6	7-5 Practice B	47
1-7 Practice B	7	7-6 Practice B	48
2-1 Practice B	8	8-1 Practice B	49
2-2 Practice B	9	8-2 Practice B	50
2-3 Practice B	10	8-3 Practice B	51
2-4 Practice B	11	8-4 Practice B	52
2-5 Practice B	12	8-5 Practice B	53
2-6 Practice B	13	8-6 Practice B	54
2-7 Practice B	14	9-1 Practice B	55
3-1 Practice B	15	9-2 Practice B	56
3-2 Practice B	16	9-3 Practice B	57
3-3 Practice B	17	9-4 Practice B	58
3-4 Practice B	18	9-5 Practice B	59
3-5 Practice B	19	9-6 Practice B	60
3-6 Practice B	20	10-1 Practice B	61
4-1 Practice B	21	10-2 Practice B	62
4-2 Practice B	22	10-3 Practice B	63
4-3 Practice B	23	10-4 Practice B	64
4-4 Practice B	24	10-5 Practice B	65
4-5 Practice B	25	10-6 Practice B	66
4-6 Practice B	26	10-7 Practice B	67
4-7 Practice B	27	10-8 Practice B	68
4-8 Practice B	28	11-1 Practice B	69
5-1 Practice B	29	11-2 Practice B	70
5-2 Practice B	30	11-3 Practice B	71
5-3 Practice B	31	11-4 Practice B	72
5-4 Practice B	32	11-5 Practice B	73
5-5 Practice B	33	11-6 Practice B	74
5-6 Practice B	34	11-7 Practice B	75
5-7 Practice B	35	12-1 Practice B	76
5-8 Practice B	36	12-2 Practice B	77
6-1 Practice B	37	12-3 Practice B	78
6-2 Practice B	38	12-4 Practice B	79
6-3 Practice B	39	12-5 Practice B	80
6-4 Practice B	40	12-6 Practice B	81
6-5 Practice B	41	12-7 Practice B	82

Holt Geometry

Name _____ Date _____ Class _____

LESSON 1-1 Practice B
Understanding Points, Lines, and Planes

Use the figure for Exercises 1–7.

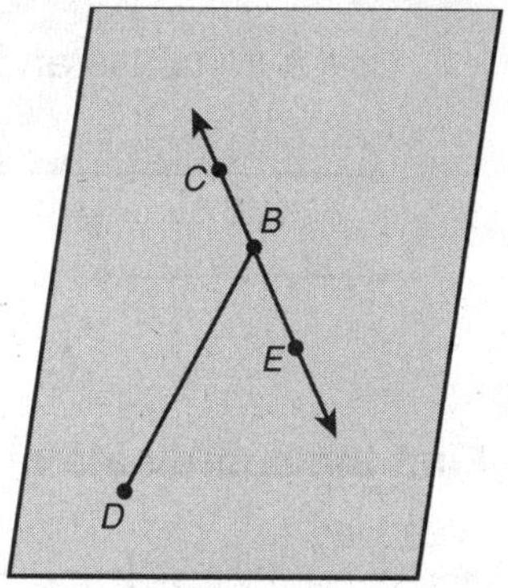

1. Name a plane. __Possible answers: plane BCD; plane BED__
2. Name a segment. __$\overline{BD}$, $\overline{BC}$, $\overline{BE}$, or $\overline{CE}$__
3. Name a line. __Possible answers: $\overleftrightarrow{EC}$; $\overleftrightarrow{BC}$; $\overleftrightarrow{BE}$__
4. Name three collinear points.
   __points B, C, and E__
5. Name three noncollinear points.
   __Possible answers: points B, C, and D or points B, E, and D__
6. Name the intersection of a line and a segment not on the line. __point B__
7. Name a pair of opposite rays. __$\overrightarrow{BC}$ and $\overrightarrow{BE}$__

Use the figure for Exercises 8–11.

8. Name the points that determine plane $\mathcal{R}$.
   __points X, Y, and Z__
9. Name the point at which line m intersects plane $\mathcal{R}$. __point Z__
10. Name two lines in plane $\mathcal{R}$ that intersect line m.
    __$\overleftrightarrow{XZ}$ and $\overleftrightarrow{YZ}$__
11. Name a line in plane $\mathcal{R}$ that does not intersect line m. __$\overleftrightarrow{XY}$__

Draw your answers in the space provided.

Michelle Kwan won a bronze medal in figure skating at the 2002 Salt Lake City Winter Olympic Games.

12. Michelle skates straight ahead from point L and stops at point M. Draw her path.

13. Michelle skates straight ahead from point L and continues through point M. Name a figure that represents her path. Draw her path. __ray__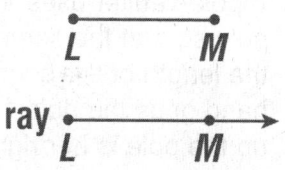

14. Michelle and her friend Alexei start back to back at point L and skate in opposite directions. Michelle skates through point M, and Alexei skates through point K. Draw their paths.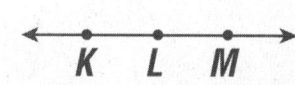

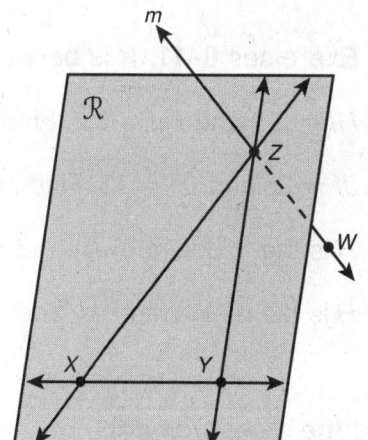

Copyright © by Holt, Rinehart and Winston.
All rights reserved.

Holt Geometry

Name _____ Date _____ Class _____

LESSON 1-2 Practice B
Measuring and Constructing Segments

Draw your answer in the space provided.

1. Use a compass and straightedge to construct $\overline{XY}$ congruent to $\overline{UV}$.

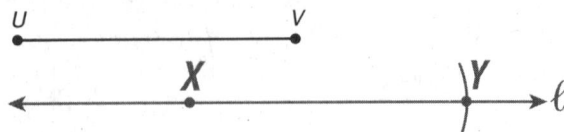

Find the coordinate of each point.

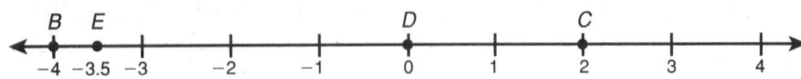

2. D __0__

3. C __2__

4. E __−3.5__

Find each length.

5. BE __0.5__

6. DB __4__

7. EC __5.5__

For Exercises 8–11, H is between I and J.

8. HI = 3.9 and HJ = 6.2. Find IJ. __10.1__

9. JI = 25 and IH = 13. Find HJ. __12__

10. H is the midpoint of $\overline{IJ}$, and IH = 0.75. Find HJ. __0.75__

11. H is the midpoint of $\overline{IJ}$, and IJ = 9.4. Find IH. __4.7__

Find the measurements.

12. K —$x + 0.5$— L —$3x - 2$— M; whole: $3x + 1.5$

Find LM. __7__

13. A pole-vaulter uses a 15-foot-long pole. She grips the pole so that the segment below her left hand is twice the length of the segment above her left hand. Her right hand grips the pole 1.5 feet above her left hand. How far up the pole is her right hand? __11.5 ft__

Name _____ Date _____ Class _____

LESSON 1-3 Practice B
Measuring and Constructing Angles

Draw your answer on the figure.

1. Use a compass and straightedge to construct angle bisector $\vec{DG}$.

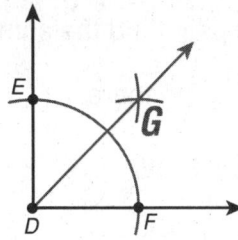

2. Name eight different angles in the figure.

   ∠A, ∠C, ∠ABC, ∠ABD, ∠ADB,

   ∠ADC, ∠CBD, and ∠CDB

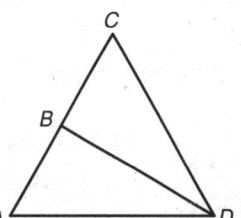

Find the measure of each angle and classify each as acute, right, obtuse, or straight.

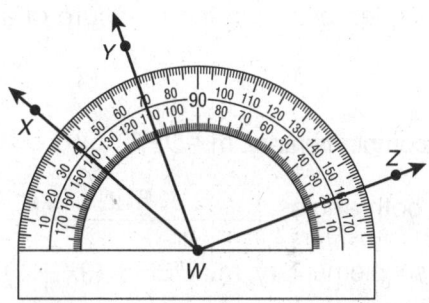

3. ∠YWZ

   90°; right

4. ∠XWZ

   120°; obtuse

5. ∠YWX

   30°; acute

T is in the interior of ∠PQR. Find each of the following.

6. m∠PQT if m∠PQR = 25° and m∠RQT = 11°.   **14°**

7. m∠PQR if m∠PQR = (10x − 7)°, m∠RQT = 5x°, and m∠PQT = (4x + 6)°.   **123°**

8. m∠PQR if $\vec{QT}$ bisects ∠PQR, m∠RQT = (10x − 13)°, and m∠PQT = (6x + 1)°.   **44°**

9. Longitude is a measurement of position around the equator of Earth. Longitude is measured in degrees, minutes, and seconds. Each degree contains 60 minutes, and each minute contains 60 seconds. Minutes are indicated by the symbol ′ and seconds are indicated by the symbol ″. Williamsburg, VA, is located at 76°42′25″. Roanoke, VA, is located at 79°57′30″. Find the difference of their longitudes in degrees, minutes, and seconds.   **3°15′05″**

10. To convert minutes and seconds into decimal parts of a degree, divide the number of minutes by 60 and the number of seconds by 3,600. Then add the numbers together. Write the location of Roanoke, VA, as a decimal to the nearest thousandths of a degree.   **79.958°**

Holt Geometry

Name _____ Date _____ Class _____

LESSON 1-4 Practice B
Pairs of Angles

1. ∠PQR and ∠SQR form a linear pair. Find the sum of their measures. __**180°**__

2. Name the ray that ∠PQR and ∠SQR share. __**$\overrightarrow{QR}$**__

Use the figures for Exercises 3 and 4.

3. supplement of ∠Z __**137.9°**__

4. complement of ∠Y __**(110 − 8x)°**__

5. An angle measures 12 degrees less than three times its supplement. Find the measure of the angle. __**132°**__

6. An angle is its own complement. Find the measure of a supplement to this angle.
__**135°**__

7. ∠DEF and ∠FEG are complementary. m∠DEF = (3x − 4)°, and m∠FEG = (5x + 6)°. Find the measures of both angles. __**m∠DEF = 29°; m∠FEG = 61°**__

8. ∠DEF and ∠FEG are supplementary. m∠DEF = (9x + 1)°, and m∠FEG = (8x + 9)°. Find the measures of both angles. __**m∠DEF = 91°; m∠FEG = 89°**__

Use the figure for Exercises 9 and 10.
In 2004, several nickels were minted to commemorate the Louisiana Purchase and Lewis and Clark's expedition into the American West. One nickel shows a pipe and a hatchet crossed to symbolize peace between the American government and Native American tribes.

9. Name a pair of vertical angles.

__**Possible answers: ∠1 and ∠3**__
__**or ∠2 and ∠4**__

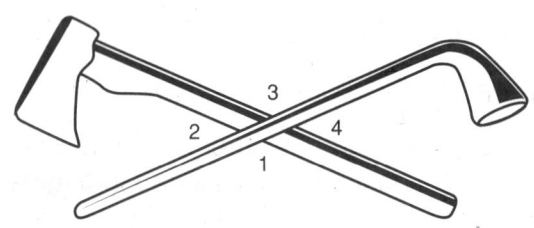

10. Name a linear pair of angles.

__**Possible answers: ∠1 and ∠2; ∠2 and ∠3; ∠3 and ∠4; or ∠1 and ∠4**__

11. ∠ABC and ∠CBD form a linear pair and have equal measures. Tell if ∠ABC is acute, right, or obtuse. __**right**__

12. ∠KLM and ∠MLN are complementary. $\overrightarrow{LM}$ bisects ∠KLN. Find the measures of ∠KLM and ∠MLN. __**45°; 45°**__

Copyright © by Holt, Rinehart and Winston.
All rights reserved.

Holt Geometry

Name _____ Date _____ Class _____

LESSON 1-5 Practice B
Using Formulas in Geometry

Use the figures for Exercises 1–3.

1. Find the perimeter of triangle A. __**12 ft**__

2. Find the area of triangle A. __**6 ft²**__

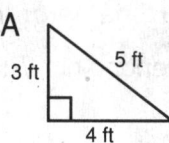

3. Triangle A is identical to triangle B. Find the height h of triangle B. __**2.4 ft or 2 2/5 ft**__

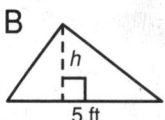

Find the perimeter and area of each shape.

4. square with a side 2.4 m in length

   __**P = 9.6 m; A = 5.76 m²**__

5. rectangle with length (x + 3) and width 7

   __**P = 2x + 20; A = 7x + 21**__

6. Although a circle does not have sides, it does have a perimeter. What is the term for the perimeter of a circle? __**circumference**__

Find the circumference and area of each circle.

7.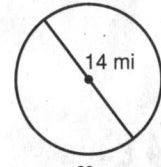

Use 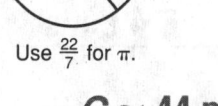 for π.

__**C ≈ 44 mi**__
__**A ≈ 154 mi²**__

8.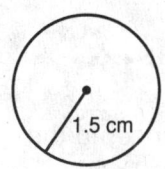

Use 3.14 for π.

__**C ≈ 9.42 cm**__
__**A ≈ 7.065 cm²**__

9.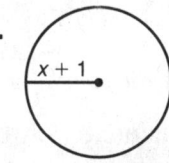

Leave π as π.

__**C ≈ 2π(x + 1)**__
__**A ≈ π(x² + 2x + 1)**__

10. The area of a square is $\frac{1}{4}$ in². Find the perimeter. __**2 in.**__

11. The area of a triangle is 152 m², and the height is 16 m. Find the base. __**19 m**__

12. The circumference of a circle is 25π mm. Find the radius. __**12.5 mm**__

Use the figure for Exercises 13 and 14.

Lucas has a 39-foot-long rope. He uses all the rope to outline this T-shape in his backyard. All the angles in the figure are right angles.

13. Find x. __**7.5 ft**__

14. Find the area enclosed by the rope. __**42 ft²**__

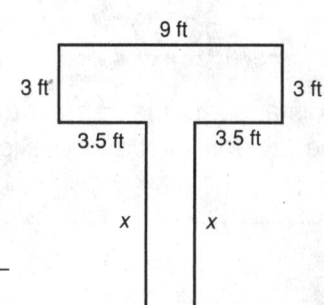

Copyright © by Holt, Rinehart and Winston.
All rights reserved.

Holt Geometry

Name _____ Date _____ Class _____

LESSON 1-6 Practice B
Midpoint and Distance in the Coordinate Plane

Find the coordinates of the midpoint of each segment.

1. $\overline{TU}$ with endpoints $T(5, -1)$ and $U(1, -5)$ __(3, −3)__

2. $\overline{VW}$ with endpoints $V(-2, -6)$ and $W(x + 2, y + 3)$ $\left(\dfrac{x}{2}, \dfrac{y-3}{2}\right)$

3. Y is the midpoint of $\overline{XZ}$. X has coordinates $(2, 4)$, and Y has coordinates $(-1, 1)$. Find the coordinates of Z.   __(−4, −2)__

Use the figure for Exercises 4–7.

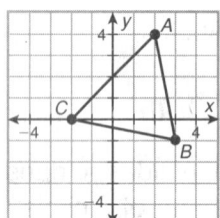

4. Find AB.   __$\sqrt{26}$ units__

5. Find BC.   __$\sqrt{26}$ units__

6. Find CA.   __$4\sqrt{2}$ units__

7. Name a pair of congruent segments.   __$\overline{AB}$ and $\overline{BC}$__

Find the distances.

8. Use the Distance Formula to find the distance, to the nearest tenth, between $K(-7, -4)$ and $L(-2, 0)$.   __6.4 units__

9. Use the Pythagorean Theorem to find the distance, to the nearest tenth, between $F(9, 5)$ and $G(-2, 2)$.   __11.4 units__

Use the figure for Exercises 10 and 11.

Snooker is a kind of pool or billiards played on a 6-foot-by-12-foot table. The side pockets are halfway down the rails (long sides).

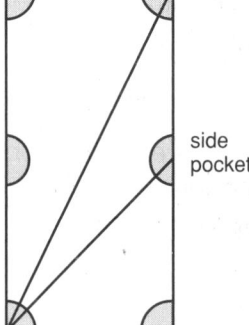

10. Find the distance, to the nearest tenth of a foot, diagonally across the table from corner pocket to corner pocket.
__13.4 ft__

11. Find the distance, to the nearest tenth of an inch, diagonally across the table from corner pocket to side pocket.
__101.8 in.__

Holt Geometry

Name _____ Date _____ Class _____

LESSON 1-7 Practice B
Transformations in the Coordinate Plane

Use the figure for Exercises 1–3.

The figure in the plane at right shows the preimage in the transformation ABCD → A'B'C'D'. Match the number of the image (below) with the name of the correct transformation.

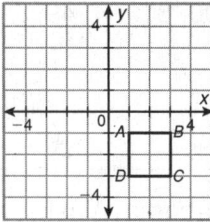

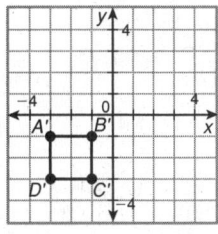 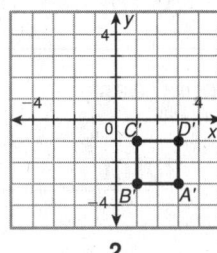 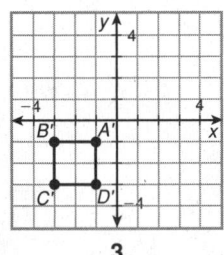

    1   2   3

1. rotation ___2___   2. translation ___1___   3. reflection ___3___

4. A figure has vertices at D(−2, 1), E(−3, 3), and F(0, 3). After a transformation, the image of the figure has vertices at D'(−1, −2), E'(−3, −3), and F'(−3, 0). Draw the preimage and the image. Then identify the transformation.

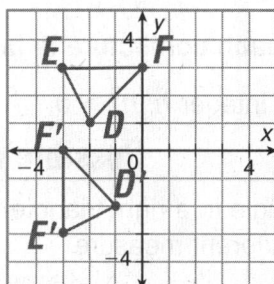

rotation

5. A figure has vertices at G(0, 0), H(−1, −2), I(−1.5, 0), and J(−2.5, 2). Find the coordinates for the image of GHIJ after the translation (x, y) → (x − 2.5, y + 4).

G'(−2.5, 4), H'(−3.5, 2), I'(−4, 4), J'(−5, 6)

Use the figure for Exercise 6.

6. A parking garage attendant will make the most money when the maximum number of cars fits in the parking garage. To fit one more car in, the attendant moves a car from position 1 to position 2. Write a rule for this translation.

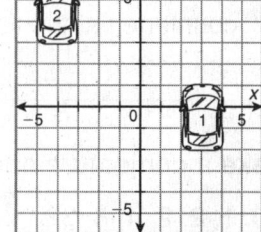

(x, y) → (x − 7, y + 5)

7. A figure has vertices at X(−1, 1), Y(−2, 3), and Z(0, 4). Draw the image of XYZ after the translation (x, y) → (x − 2, y) and a 180° rotation around X.

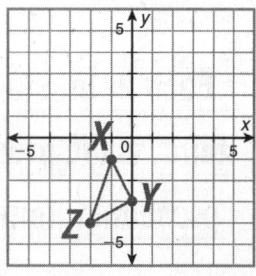

Copyright © by Holt, Rinehart and Winston.
All rights reserved.

Holt Geometry

Name _____ Date _____ Class _____

LESSON 2-1 Practice B
Using Inductive Reasoning to Make Conjectures

Find the next item in each pattern.

1. 100, 81, 64, 49, . . .

   _____36_____

2. 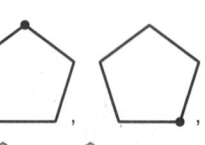, . . .

3. Alabama, Alaska, Arizona, . . .

   ___Arkansas___

4. west, south, east, . . .

   ___north___

Complete each conjecture.

5. The square of any negative number is ___positive___.

6. The number of segments determined by n points is ___$\dfrac{n(n-1)}{2}$___.

Show that each conjecture is false by finding a counterexample.

7. For any integer n, $n^3 > 0$.

   ___Possible answers: zero, any negative number___

8. Each angle in a right triangle has a different measure.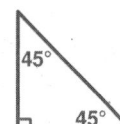

9. For many years in the United States, each bank printed its own currency. The variety of different bills led to widespread counterfeiting. By the time of the Civil War, a significant fraction of the currency in circulation was counterfeit. If one Civil War soldier had 48 bills, 16 of which were counterfeit, and another soldier had 39 bills, 13 of which were counterfeit, make a conjecture about what fraction of bills were counterfeit at the time of the Civil War.

   ___One-third of the bills were counterfeit.___

Make a conjecture about each pattern. Write the next two items.

10. 1, 2, 2, 4, 8, 32, . . .

    ___Each item, starting with the third, is the product of the two preceding items; 256, 8192.___

11. , . . .

    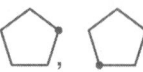

    ___The dot skips over one vertex in a clockwise direction.___

Holt Geometry

Practice B
LESSON 2-2 Conditional Statements

Identify the hypothesis and conclusion of each conditional.

1. If you can see the stars, then it is night.
   Hypothesis: **You can see the stars.**
   Conclusion: **It is night.**

2. A pencil writes well if it is sharp.
   Hypothesis: **A pencil is sharp.**
   Conclusion: **The pencil writes well.**

Write a conditional statement from each of the following.

3. Three noncollinear points determine a plane.

   If three points are noncollinear, then they determine a plane.

4. 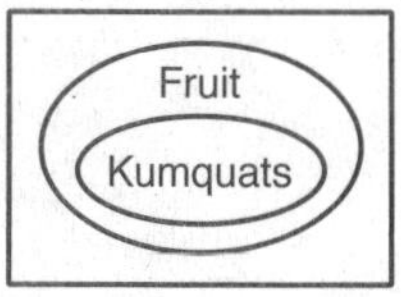 (Fruit / Kumquats diagram) **If a food is a kumquat, then it is a fruit.**

Determine if each conditional is true. If false, give a counterexample.

5. If two points are noncollinear, then a right triangle contains one obtuse angle.

   true

6. If a liquid is water, then it is composed of hydrogen and oxygen.

   true

7. If a living thing is green, then it is a plant.

   false; sample answer: a frog

8. "If G is at 4, then GH is 3." Write the converse, inverse, and contrapositive of this statement. Find the truth value of each.

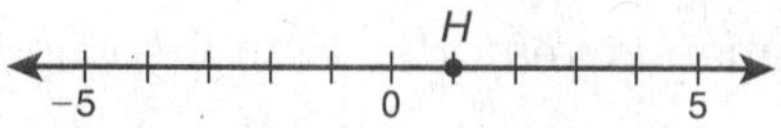

   Converse: **If GH is 3, then G is at 4; false**
   Inverse: **If G is not at 4, then GH is not 3; false**
   Contrapositive: **If GH is not 3, then G is not at 4; true**

This chart shows a small part of the *Mammalia* class of animals, the mammals. Write a conditional to describe the relationship between each given pair.

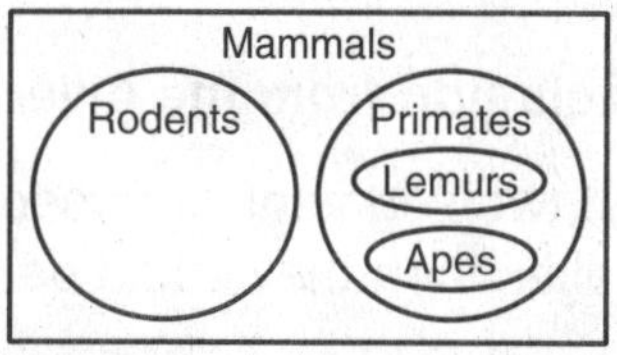

9. primates and mammals **If an animal is a primate, then it is a mammal.**

10. lemurs and rodents **Sample answer: If an animal is a lemur, then it is not a rodent.**

11. rodents and apes **Sample answer: If an animal is a rodent, then it is not an ape.**

12. apes and mammals **If an animal is an ape, then it is a mammal.**

Name _____ Date _____ Class _____

LESSON 2-3 Practice B
Using Deductive Reasoning to Verify Conjectures

Tell whether each conclusion is a result of inductive or deductive reasoning.

1. The United States Census Bureau collects data on the earnings of American citizens. Using data for the three years from 2001 to 2003, the bureau concluded that the national average median income for a four-person family was $43,527.

   inductive reasoning

2. A speeding ticket costs $40 plus $5 per mi/h over the speed limit. Lynne mentions to Frank that she was given a ticket for $75. Frank concludes that Lynne was driving 7 mi/h over the speed limit.

   deductive reasoning

Determine if each conjecture is valid by the Law of Detachment.

3. Given: If m∠ABC = m∠CBD, then $\overrightarrow{BC}$ bisects ∠ABD. $\overrightarrow{BC}$ bisects ∠ABD.
   Conjecture: m∠ABC = m∠CBD.   **invalid**

4. Given: You will catch a catfish if you use stink bait. Stuart caught a catfish.
   Conjecture: Stuart used stink bait.   **invalid**

5. Given: An obtuse triangle has two acute angles. Triangle ABC is obtuse.
   Conjecture: Triangle ABC has two acute angles.   **valid**

Determine if each conjecture is valid by the Law of Syllogism.

6. Given: If the gossip said it, then it must be true. If it is true, then somebody is in big trouble.
   Conjecture: Somebody is in big trouble because the gossip said it.   **valid**

7. Given: No human is immortal. Fido the dog is not human.
   Conjecture: Fido the dog is immortal.   **invalid**

8. Given: The radio is distracting when I am studying. If it is 7:30 P.M. on a weeknight, I am studying.
   Conjecture: If it is 7:30 P.M. on a weeknight, the radio is distracting.   **valid**

Draw a conclusion from the given information.

9. Given: If two segments intersect, then they are not parallel. If two segments are not parallel, then they could be perpendicular. $\overline{EF}$ and $\overline{MN}$ intersect.

   $\overline{EF}$ and $\overline{MN}$ could be perpendicular.

10. Given: When you are relaxed, your blood pressure is relatively low. If you are sailing, you are relaxed. Becky is sailing.

    Becky's blood pressure is relatively low.

Name _____ Date _____ Class _____

LESSON 2-4 Practice B
Biconditional Statements and Definitions

Write the conditional statement and converse within each biconditional.

1. The tea kettle is whistling if and only if the water is boiling.

   Conditional: **If the tea kettle is whistling, then the water is boiling.**

   Converse: **If the water is boiling, then the tea kettle is whistling.**

2. A biconditional is true if and only if the conditional and converse are both true.

   Conditional: **If a biconditional is true, then the conditional and converse are both true.**

   Converse: **If the conditional and converse are both true, then the biconditional is true.**

For each conditional, write the converse and a biconditional statement.

3. Conditional: If n is an odd number, then $n - 1$ is divisible by 2.

   Converse: **If $n - 1$ is divisible by 2, then n is an odd number.**

   Biconditional: **n is an odd number if and only if $n - 1$ is divisible by 2.**

4. Conditional: An angle is obtuse when it measures between 90° and 180°.

   Converse: **If an angle measures between 90° and 180°, then the angle is obtuse.**

   Biconditional: **An angle is obtuse if and only if it measures between 90° and 180°.**

Determine whether a true biconditional can be written from each conditional statement. If not, give a counterexample.

5. If the lamp is unplugged, then the bulb does not shine.

   No; sample answer: The switch could be off.

6. The date can be the 29th if and only if it is not February.

   No; possible answer: Leap years have a Feb. 29th.

Write each definition as a biconditional.

7. A cube is a three-dimensional solid with six square faces.

   A figure is a cube if and only if it is a three-dimensional solid with six square faces.

8. Tanya claims that the definition of *doofus* is "her younger brother."

   A person is a doofus if and only if the person is Tanya's younger brother.

Name _____ Date _____ Class _____

LESSON 2-5 Practice B
Algebraic Proof

Solve each equation. Show all your steps and write a justification for each step.

1. $\frac{1}{5}(a + 10) = -3$

   $5\left[\frac{1}{5}(a + 10)\right] = 5(-3)$ (Mult. Prop. of =)
   $a + 10 = -15$ (Simplify.)
   $a + 10 - 10 = -15 - 10$ (Subtr. Prop. of =)
   $a = -25$ (Simplify.)

2. $t + 6.5 = 3t - 1.3$

   $t + 6.5 - t = 3t - 1.3 - t$ (Subtr. Prop. of =)
   $6.5 = 2t - 1.3$ (Simplify.)
   $6.5 + 1.3 = 2t - 1.3 + 1.3$ (Add. Prop. of =)
   $7.8 = 2t$ (Simplify.)
   $\frac{7.8}{2} = \frac{2t}{2}$ (Div. Prop. of =)
   $3.9 = t$ (Simplify.)
   $t = 3.9$ (Symmetric Prop. of =)

3. The formula for the perimeter P of a rectangle with length ℓ and width w is $P = 2(\ell + w)$. Find the length of the rectangle shown here if the perimeter is $9\frac{1}{2}$ feet. Solve the equation for ℓ and justify each step. **Possible answer:**

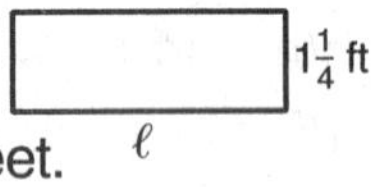

   $P = 2(\ell + w)$ (Given)
   $9\frac{1}{2} = 2(\ell + 1\frac{1}{4})$ (Subst. Prop. of =)
   $9\frac{1}{2} = 2\ell + 2\frac{1}{2}$ (Distrib. Prop.)
   $9\frac{1}{2} - 2\frac{1}{2} = 2\ell + 2\frac{1}{2} - 2\frac{1}{2}$ (Subtr. Prop. of =)

   $7 = 2\ell$ (Simplify.)
   $\frac{7}{2} = \frac{2\ell}{2}$ (Div. Prop. of =)
   $3\frac{1}{2} = \ell$ (Simplify.)
   $\ell = 3\frac{1}{2}$ (Symmetric Prop. of =)

Write a justification for each step.

4. 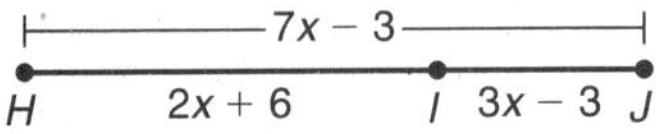
   $HJ = HI + IJ$ __Seg. Add. Post.__
   $7x - 3 = (2x + 6) + (3x - 3)$ __Subst. Prop. of =__
   $7x - 3 = 5x + 3$ __Simplify.__
   $7x = 5x + 6$ __Add. Prop. of =__
   $2x = 6$ __Subtr. Prop. of =__
   $x = 3$ __Div. Prop. of =__

Identify the property that justifies each statement.

5. $m = n$, so $n = m$.

   __Symmetric Prop. of =__

6. $\angle ABC \cong \angle ABC$

   __Reflexive Prop. of $\cong$__

7. $\overline{KL} \cong \overline{LK}$

   __Reflexive Prop. of $\cong$__

8. $p = q$ and $q = -1$, so $p = -1$.

   __Transitive Prop. of = or Subst.__

Name _____ Date _____ Class _____

LESSON 2-6
Practice B
Geometric Proof

Write a justification for each step.

Given: AB = EF, B is the midpoint of $\overline{AC}$, and E is the midpoint of $\overline{DF}$.

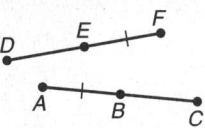

1. B is the midpoint of $\overline{AC}$, and E is the midpoint of $\overline{DF}$. — Given
2. $\overline{AB} \cong \overline{BC}$, and $\overline{DE} \cong \overline{EF}$. — Def. of mdpt.
3. AB = BC, and DE = EF. — Def. of ≅ segments
4. AB + BC = AC, and DE + EF = DF. — Seg. Add. Post.
5. 2AB = AC, and 2EF = DF. — Subst.
6. AB = EF — Given
7. 2AB = 2EF — Mult. Prop. of =
8. AC = DF — Subst. Prop. of =
9. $\overline{AC} \cong \overline{DF}$ — Def. of ≅ segments

Fill in the blanks to complete the two-column proof.

10. **Given:** ∠HKJ is a straight angle.
    $\vec{KI}$ bisects ∠HKJ.
    Prove: ∠IKJ is a right angle.

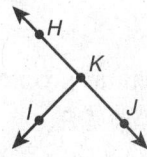

Proof:

Statements	Reasons
1. a. __∠HKJ is a straight angle.__	1. Given
2. m∠HKJ = 180°	2. b. __Def. of straight ∠__
3. c. __$\vec{KI}$ bisects ∠HKJ__	3. Given
4. ∠IKJ ≅ ∠IKH	4. Def. of ∠ bisector
5. m∠IKJ = m∠IKH	5. Def. of ≅
6. d. __m∠IKJ + m∠IKH = m∠HKJ__	6. ∠ Add. Post.
7. 2m∠IKJ = 180°	7. e. Subst. (Steps __2, 5, 6__)
8. m∠IKJ = 90°	8. Div. Prop. of =
9. ∠IKJ is a right angle.	9. f. __Def. of right ∠__

Copyright © by Holt, Rinehart and Winston.
All rights reserved.

13

Holt Geometry

Name _____ Date _____ Class _____

LESSON 2-7 Practice B
Flowchart and Paragraph Proofs

1. Use the given two-column proof to write a flowchart proof.
   Given: ∠4 ≅ ∠3
   Prove: m∠1 = m∠2

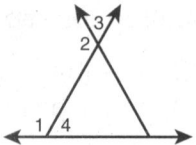

Statements	Reasons
1. ∠1 and ∠4 are supplementary, ∠2 and ∠3 are supplementary.	1. Linear Pair Thm.
2. ∠4 ≅ ∠3	2. Given
3. ∠1 ≅ ∠2	3. ≅ Supps. Thm.
4. m∠1 = m∠2	4. Def. of ≅ ∕s

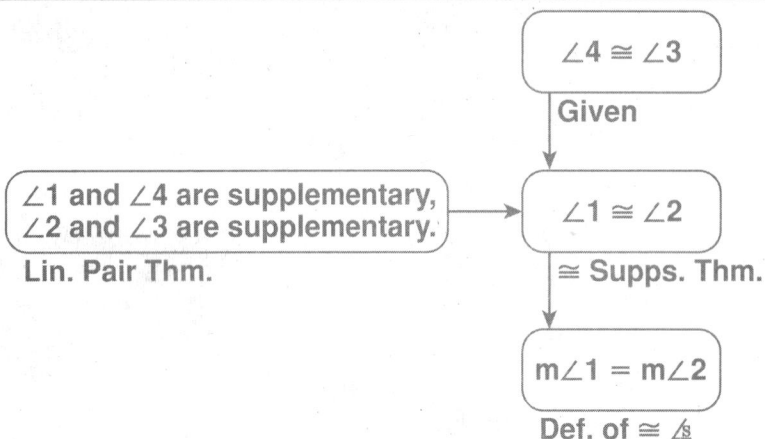

2. Use the given two-column proof to write a paragraph proof.
   Given: AB = CD, BC = DE
   Prove: C is the midpoint of $\overline{AE}$.

Statements	Reasons
1. AB = CD, BC = DE	1. Given
2. AB + BC = CD + DE	2. Add. Prop. of =
3. AB + BC = AC, CD + DE = CE	3. Seg. Add. Post.
4. AC = CE	4. Subst.
5. $\overline{AC}$ ≅ $\overline{CE}$	5. Def. of ≅ segs.
6. C is the midpoint of $\overline{AE}$.	6. Def. of mdpt.

It is given that AB = CD and BC = DE, so by the Addition Property of Equality, AB + BC = CD + DE. But by the Segment Addition Postulate, AB + BC = AC and CD + DE = CE. Therefore substitution yields AC = CE. By the definition of congruent segments, $\overline{AC}$ ≅ $\overline{CE}$ and thus C is the midpoint of $\overline{AE}$ by the definition of midpoint.

Name _____ Date _____ Class _____

LESSON 3-1 Practice B
Lines and Angles

For Exercises 1–4, identify each of the following in the figure. **Sample answers:**

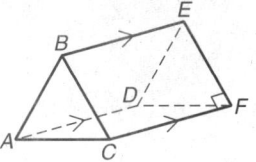

1. a pair of parallel segments $\overline{BE} \parallel \overline{AD}$

2. a pair of skew segments $\overline{AB}$ and $\overline{CF}$ are skew.

3. a pair of perpendicular segments $\overline{CF} \perp \overline{EF}$

4. a pair of parallel planes   plane $ABC \parallel$ plane DEF

In Exercises 5–10, give one example of each from the figure.

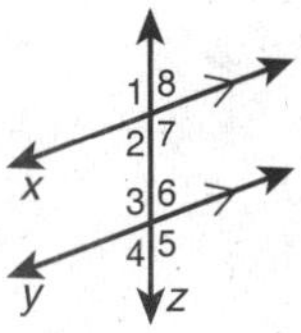

5. a transversal

   line z

6. parallel lines

   lines x and y

7. corresponding angles

   Sample answer:
   $\angle 1$ and $\angle 3$

8. alternate interior angles

   Sample answer:
   $\angle 2$ and $\angle 6$

9. alternate exterior angles

   Sample answer:
   $\angle 1$ and $\angle 5$

10. same-side interior angles

    Sample answer:
    $\angle 2$ and $\angle 3$

Use the figure for Exercises 11–14. The figure shows a utility pole with an electrical line and a telephone line. The angled wire is a tension wire. For each angle pair given, identify the transversal and classify the angle pair. (*Hint:* Think of the utility pole as a line for these problems.)

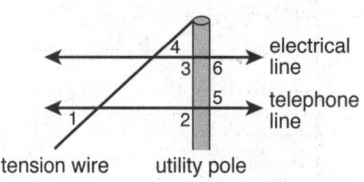

11. $\angle 5$ and $\angle 6$

    transv.: utility pole; same-side interior angles

12. $\angle 1$ and $\angle 4$

    transv.: tension wire; alternate exterior angles

13. $\angle 1$ and $\angle 2$

    transv.: telephone line; corresponding angles

14. $\angle 5$ and $\angle 3$

    transv.: utility pole; alternate interior angles

Name _____ Date _____ Class _____

LESSON 3-2 Practice B
Angles Formed by Parallel Lines and Transversals

Find each angle measure.

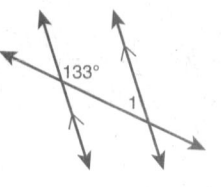

1. m∠1 _____47°_____

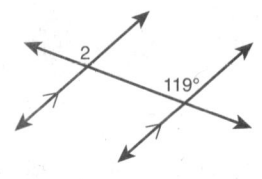

2. m∠2 _____119°_____

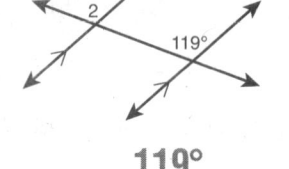

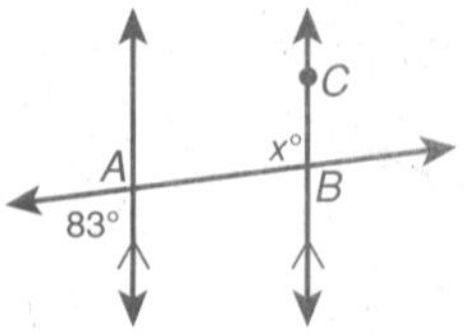

3. m∠ABC _____97°_____

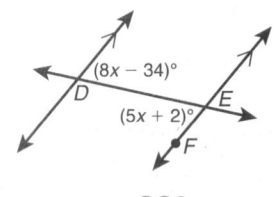

4. m∠DEF _____62°_____

Complete the two-column proof to show that same-side exterior angles are supplementary.

5. **Given:** $p \parallel q$

   Prove: m∠1 + m∠3 = 180°

   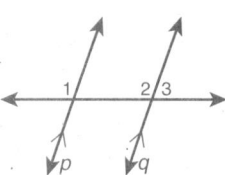

   Proof:

Statements	Reasons
1. $p \parallel q$	1. Given
2. a. __m∠2 + m∠3 = 180°__	2. Lin. Pair Thm.
3. ∠1 ≅ ∠2	3. b. __Corr. ∠s Post.__
4. c. __m∠1 = m∠2__	4. Def. of ≅ ∠s
5. d. __m∠1 + m∠3 = 180°__	5. e. __Subst.__

6. Ocean waves move in parallel lines toward the shore. The figure shows Sandy Beaches windsurfing across several waves. For this exercise, think of Sandy's wake as a line. m∠1 = (2x + 2y)° and m∠2 = (2x + y)°. Find x and y.

   x = ___15___

   y = ___40___

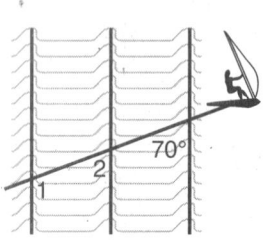

Copyright © by Holt, Rinehart and Winston.
All rights reserved.

Holt Geometry

Name _____ Date _____ Class _____

LESSON 3-3 Practice B
Proving Lines Parallel

Use the figure for Exercises 1–8. Tell whether lines *m* and *n* must be parallel from the given information. If they are, state your reasoning. (*Hint:* The angle measures may change for each exercise, and the figure is for reference only.)

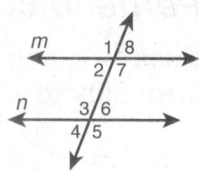

1. ∠7 ≅ ∠3

   m ∥ *n*; Conv. of Alt. Int. ∠s Thm.

2. m∠3 = (15x + 22)°, m∠1 = (19x − 10)°, x = 8

   m ∥ *n*; Conv. of Corr. ∠s Post.

3. ∠7 ≅ ∠6

   m and *n* are parallel if and only if m∠7 = 90°.

4. m∠2 = (5x + 3)°, m∠3 = (8x − 5)°, x = 14

   m ∥ *n*; Conv. of Same-Side Int. ∠s Thm.

5. m∠8 = (6x − 1)°, m∠4 = (5x + 3)°, x = 9

   m and *n* are not parallel.

6. ∠5 ≅ ∠7

   m ∥ *n*; Conv. of Corr. ∠s Post.

7. ∠1 ≅ ∠5

   m ∥ *n*; Conv. of Alt. Ext. ∠s Thm.

8. m∠6 = (x + 10)°, m∠2 = (x + 15)°

   m and *n* are not parallel.

9. Look at some of the printed letters in a textbook. The small horizontal and vertical segments attached to the ends of the letters are called *serifs*. Most of the letters in a textbook are in a serif typeface. The letters on this page do not have serifs, so these letters are in a sans-serif typeface. (*Sans* means "without" in French.) The figure shows a capital letter *A* with serifs. Use the given information to write a paragraph proof that the serif, segment $\overline{HI}$, is parallel to segment $\overline{JK}$.

   Given: ∠1 and ∠3 are supplementary.

   Prove: $\overline{HI} \parallel \overline{JK}$

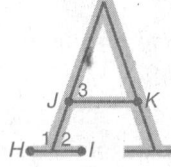

Sample answer: The given information states that ∠1 and ∠3 are supplementary. ∠1 and ∠2 are also supplementary by the Linear Pair Theorem. Therefore ∠3 and ∠2 must be congruent by the Congruent Supplements Theorem. Since ∠3 and ∠2 are congruent, $\overline{HI}$ and $\overline{JK}$ are parallel by the Converse of the Corresponding Angles Postulate.

Copyright © by Holt, Rinehart and Winston.
All rights reserved.

Holt Geometry

Name _____ Date _____ Class _____

LESSON 3-4
Practice B
Perpendicular Lines

For Exercises 1–4, name the shortest segment from the point to the line and write an inequality for *x*. (*Hint:* One answer is a double inequality.)

1.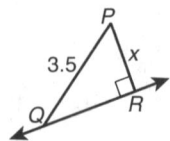

   $\overline{PR}$; $x < 3.5$

2.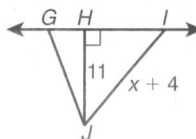

   $\overline{HJ}$; $x > 7$

3.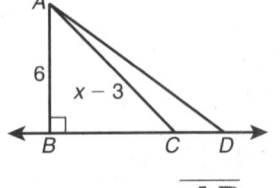

   $\overline{AB}$; $x > 9$

4.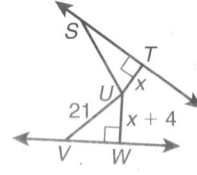

   $\overline{UT}$; $x < 17$

Complete the two-column proof.

5. **Given:** $m \perp n$
   Prove: ∠1 and ∠2 are a linear pair of congruent angles.
   Proof:

   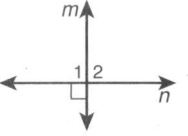

Statements	Reasons
1. a. _____ $m \perp n$ _____	1. Given
2. b. $m\angle 1 = 90°$, $m\angle 2 = 90°$	2. Def. of $\perp$
3. ∠1 ≅ ∠2	3. c. _____ Def. of ≅ ∠s _____
4. $m\angle 1 + m\angle 2 = 180°$	4. Add. Prop. of =
5. d. _∠1 and ∠2 are a linear pair._	5. Def. of linear pair

6. The Four Corners National Monument is at the intersection of the borders of Arizona, Colorado, New Mexico, and Utah. It is called the four corners because the intersecting borders are perpendicular. If you were to lie down on the intersection, you could be in four states at the same time—the only place in the United States where this is possible. The figure shows the Colorado-Utah border extending north in a straight line until it intersects the Wyoming border at a right angle. Explain why the Colorado-Wyoming border must be parallel to the Colorado–New Mexico border. **Possible answer:**

   All the borders are straight lines, and the Colorado–Utah border is a transversal to the Colorado–Wyoming and the Colorado–New Mexico borders. Because the transversal is perpendicular to both borders, the borders must be parallel.

Copyright © by Holt, Rinehart and Winston.
All rights reserved.

Holt Geometry

Name _____ Date _____ Class _____

LESSON 3-5 Practice B
Slopes of Lines

Use the slope formula to determine the slope of each line.

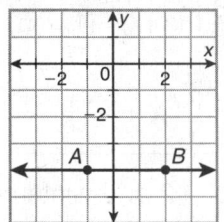

1. $\overleftrightarrow{AB}$ __**zero**__

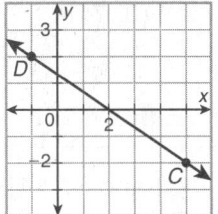

2. $\overleftrightarrow{CD}$ __$-\dfrac{2}{3}$__

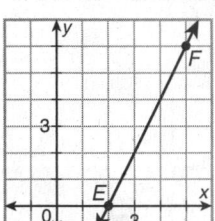

3. $\overleftrightarrow{EF}$ __**2**__

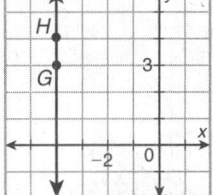

4. $\overleftrightarrow{GH}$ __**undefined**__

Graph each pair of lines. Use slopes to determine whether the lines are parallel, perpendicular, or neither.

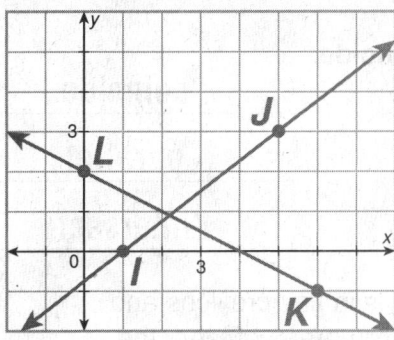

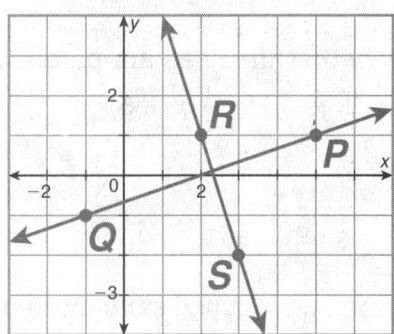

5. $\overleftrightarrow{IJ}$ and $\overleftrightarrow{KL}$ for $I(1, 0)$, $J(5, 3)$, $K(6, -1)$, and $L(0, 2)$ __**neither**__

6. $\overleftrightarrow{PQ}$ and $\overleftrightarrow{RS}$ for $P(5, 1)$, $Q(-1, -1)$, $R(2, 1)$, and $S(3, -2)$ __**perpendicular**__

7. At a ski resort, the different ski runs down the mountain are color-coded according to difficulty. Green is easy, blue is medium, and black is hard. Assume that the ski runs below are rated only according to their slope (steeper is harder) and that there is one green, one blue, and one black run. Assign a color to each ski run.

Emerald $\left(m = \dfrac{4}{7}\right)$ __**green**__

Diamond $\left(m = \dfrac{5}{4}\right)$ __**black**__

Ruby $\left(m = \dfrac{5}{8}\right)$ __**blue**__

Name _____ Date _____ Class _____

LESSON 3-6 Practice B
Lines in the Coordinate Plane

Write the equation of each line in the given form.

1. the horizontal line through (3, 7) in point-slope form

   $y - 7 = 0$

2. the line with slope $-\frac{8}{5}$ through (1, −5) in point-slope form

   $y + 5 = -\frac{8}{5}(x - 1)$

3. the line through $\left(-\frac{1}{2}, -\frac{7}{2}\right)$ and (2, 14) in slope-intercept form

   $y = 7x$

4. the line with x-intercept −2 and y-intercept −1 in slope-intercept form

   $y = -\frac{1}{2}x - 1$

Graph each line.

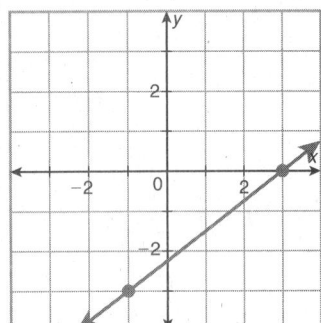

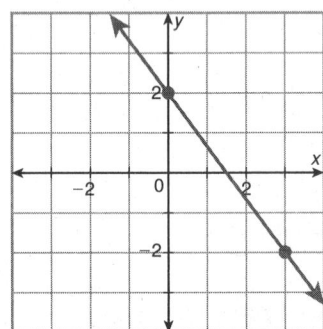

5. $y + 3 = \frac{3}{4}(x + 1)$

6. $y = -\frac{4}{3}x + 2$

Determine whether the lines are parallel, intersect, or coincide.

7. $x - 5y = 0$, $y + 1 = \frac{1}{5}(x + 5)$ _____**coincide**_____

8. $2y + 2 = x$, $\frac{1}{2}x = -1 + y$ _____**parallel**_____

9. $y = 4(x - 3)$, $\frac{3}{4} + 4y = -\frac{1}{4}x$ _____**intersect**_____

An *aquifer* is an underground storehouse of water. The water is in tiny crevices and pockets in the rock or sand, but because aquifers underlay large areas of land, the amount of water in an aquifer can be vast. Wells and springs draw water from aquifers.

10. Two relatively small aquifers are the Rush Springs (RS) aquifer and the Arbuckle-Simpson (AS) aquifer, both in Oklahoma. Suppose that starting on a certain day in 1985, 52 million gallons of water per day were taken from the RS aquifer, and 8 million gallons of water per day were taken from the AS aquifer. If the RS aquifer began with 4500 million gallons of water and the AS aquifer began with 3000 million gallons of water and no rain fell, write a slope-intercept equation for each aquifer and find how many days passed until both aquifers held the same amount of water. (Round to the nearest day.)

    RS: $y = -52x + 4500$; AS: $y = -8x + 3000$; 34 days

Name _____ Date _____ Class _____

LESSON 4-1 Practice B
Classifying Triangles

Classify each triangle by its angle measures.
(*Note:* Some triangles may belong to more than one class.)

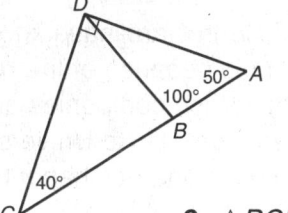

1. △ABD

   __obtuse__

2. △ADC

   __right__

3. △BCD

   __acute__

Classify each triangle by its side lengths.
(*Note:* Some triangles may belong to more than one class.)

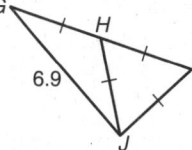

4. △GIJ

   __scalene__

5. △HIJ

   __equilateral; isosceles__

6. △GHJ

   __isosceles__

Find the side lengths of each triangle.

7.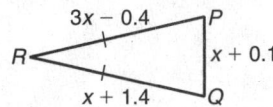

   PR = RQ = 2.3; PQ = 1

8.

   ST = SU = TU = $5\frac{1}{4}$

9. Min works in the kitchen of a catering company. Today her job is to cut whole pita bread into small triangles. Min uses a cutting machine, so every pita triangle comes out the same. The figure shows an example. Min has been told to cut 3 pita triangles for every guest. There will be 250 guests. If the pita bread she uses comes in squares with 20-centimeter sides and she doesn't waste any bread, how many squares of whole pita bread will Min have to cut up?

   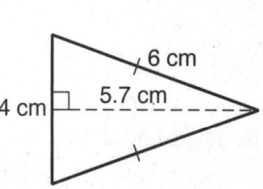

   22 pieces of pita bread

10. Follow these instructions and use a protractor to draw a triangle with sides of 3 cm, 4 cm, and 5 cm. First draw a 5-cm segment. Set your compass to 3 cm and make an arc from one end of the 5-cm segment. Now set your compass to 4 cm and make an arc from the other end of the 5-cm segment. Mark the point where the arcs intersect. Connect this point to the ends of the 5-cm segment. Classify the triangle by sides and by angles. Use the Pythagorean Theorem to check your answer.

    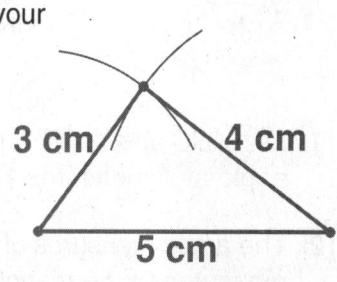

    scalene, right

Copyright © by Holt, Rinehart and Winston.
All rights reserved.

Holt Geometry

Name _____ Date _____ Class _____

LESSON 4-2 Practice B
Angle Relationships in Triangles

1. An area in central North Carolina is known as the Research Triangle because of the relatively large number of high-tech companies and research universities located there. Duke University, the University of North Carolina at Chapel Hill, and North Carolina State University are all within this area. The Research Triangle is roughly bounded by the cities of Chapel Hill, Durham, and Raleigh. From Chapel Hill, the angle between Durham and Raleigh measures 54.8°. From Raleigh, the angle between Chapel Hill and Durham measures 24.1°. Find the angle between Chapel Hill and Raleigh from Durham.   __101.1°__

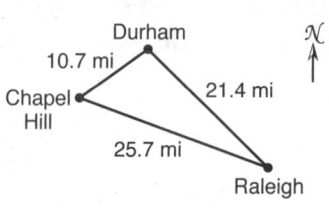

2. The acute angles of right triangle ABC are congruent. Find their measures.   __45°__

The measure of one of the acute angles in a right triangle is given. Find the measure of the other acute angle.

3. 44.9°   __45.1°__   4. (90 − z)°   __z°__   5. 0.3°   __89.7°__

Find each angle measure.

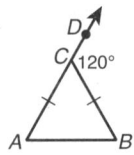
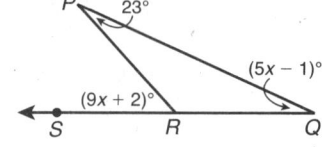

6. m∠B   __60°__   7. m∠PRS   __47°__

8. In △LMN, the measure of an exterior angle at N measures 99°. m∠L = $\frac{1}{3}x°$ and m∠M = $\frac{2}{3}x°$. Find m∠L, m∠M, and m∠LNM.   __33°; 66°; 81°__

9. m∠E and m∠G   __44°; 44°__   10. m∠T and m∠V   __108°; 108°__

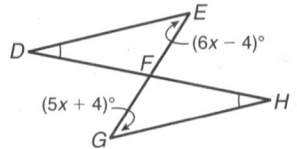
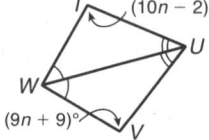

11. In △ABC and △DEF, m∠A = m∠D and m∠B = m∠E. Find m∠F if an exterior angle at A measures 107°, m∠B = (5x + 2)°, and m∠C = (5x + 5)°.   __55°__

12. The angle measures of a triangle are in the ratio 3 : 4 : 3. Find the angle measures of the triangle.   __54°; 72°; 54°__

LESSON 4-3 Practice B
Congruent Triangles

In baseball, home plate is a pentagon. Pentagon *ABCDE* is a diagram of a regulation home plate. The baseball rules are very specific about the exact dimensions of this pentagon so that every home plate is congruent to every other home plate. If pentagon *PQRST* is another home plate, identify each congruent corresponding part.

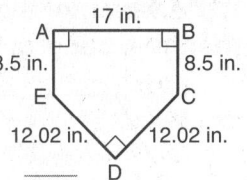

1. ∠S ≅ __∠D__
2. ∠B ≅ __∠Q__
3. $\overline{EA}$ ≅ __$\overline{TP}$__
4. ∠E ≅ __∠T__
5. $\overline{PQ}$ ≅ __$\overline{AB}$__
6. $\overline{TS}$ ≅ __$\overline{ED}$__

Given: △DEF ≅ △LMN. Find each value.

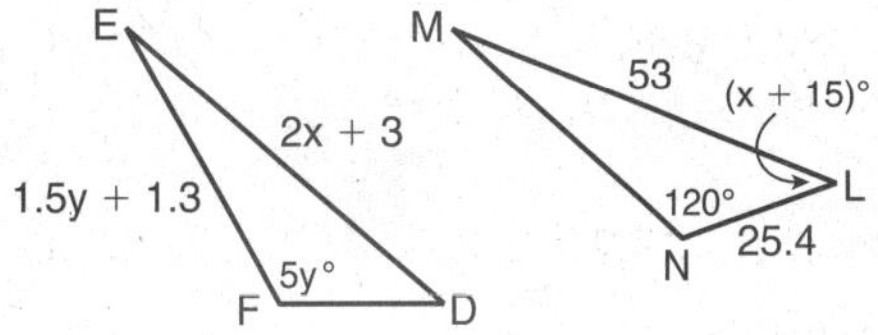

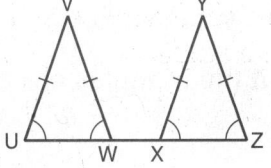

7. m∠L = __40°__
8. EF = __37.3__

9. Write a two-column proof.

Given: ∠U ≅ ∠UWV ≅ ∠ZXY ≅ ∠Z,
$\overline{UV}$ ≅ $\overline{WV}$ ≅ $\overline{XY}$ ≅ $\overline{ZY}$, $\overline{UX}$ ≅ $\overline{WZ}$

Prove: △UVW ≅ △XYZ

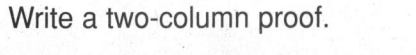

Proof: **Possible answer:**

Statements	Reasons
1. ∠U ≅ ∠UWV ≅ ∠ZXY ≅ ∠Z	1. Given
2. ∠V ≅ ∠Y	2. Third ∠s Thm.
3. $\overline{UV}$ ≅ $\overline{WV}$, $\overline{XY}$ ≅ $\overline{ZY}$	3. Given
4. $\overline{UX}$ ≅ $\overline{WZ}$	4. Given
5. UX = WZ, WX = WX	5. Def. of ≅ segs. Reflexive Prop. of =
6. UX = UW + WX, WZ = XZ + WX	6. Seg. Add. Post.
7. UW + WX = XZ + WX	7. Subst.
8. UW = XZ	8. Subtr. Prop. of =
9. △UVW ≅ △XYZ	9. Def. of ≅ △s

10. Given: △CDE ≅ △HIJ, DE = 9x, and IJ = 7x + 3. Find x and DE.

$x = \frac{3}{2}$; $DE = 13\frac{1}{2}$

11. Given: △CDE ≅ △HIJ, m∠D = (5y + 1)°, and m∠I = (6y − 25)°.
    Find y and m∠D.

    y = 26; m∠D = 131°

Name _____ Date _____ Class _____

LESSON 4-4
Practice B
Triangle Congruence: SSS and SAS

Write which of the SSS or SAS postulates, if either, can be used to prove the triangles congruent. If no triangles can be proved congruent, write *neither*.

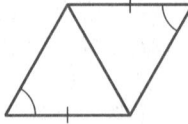

1. _____neither_____

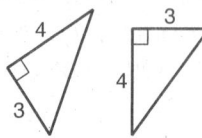

2. _____SAS_____

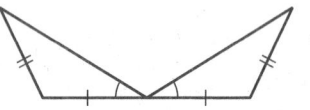

3. _____neither_____

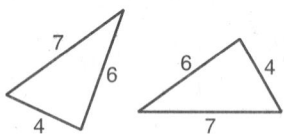

4. _____SSS_____

Find the value of *x* so that the triangles are congruent.

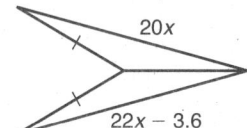

5. *x* = _____1.8_____

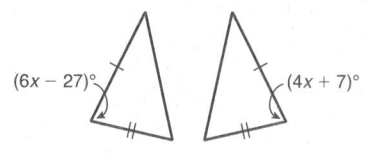

6. *x* = _____17_____

The Hatfield and McCoy families are feuding over some land. Neither family will be satisfied unless the two triangular fields are exactly the same size. You know that *C* is the midpoint of each of the intersecting segments. Write a two-column proof that will settle the dispute.

7. **Given:** *C* is the midpoint of $\overline{AD}$ and $\overline{BE}$.

    Prove: △*ABC* ≅ △*DEC*

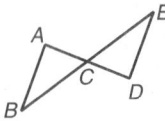

Proof: Possible answer:

Statements	Reasons
1. *C* is the midpoint of $\overline{AD}$ and $\overline{BE}$.	1. Given
2. *AC* = *CD*, *BC* = *CE*	2. Def. of mdpt.
3. $\overline{AC}$ ≅ $\overline{CD}$, $\overline{BC}$ ≅ $\overline{CE}$	3. Def. of ≅ segs.
4. ∠*ACB* ≅ ∠*DCE*	4. Vert. ∠s Thm.
5. △*ABC* ≅ △*DEC*	5. SAS

Copyright © by Holt, Rinehart and Winston.
All rights reserved.

Holt Geometry

Name _____ Date _____ Class _____

LESSON 4-5 Practice B
Triangle Congruence: ASA, AAS, and HL

Students in Mrs. Marquez's class are watching a film on the uses of geometry in architecture. The film projector casts the image on a flat screen as shown in the figure. The dotted line is the bisector of ∠ABC. Tell whether you can use each congruence theorem to prove that △ABD ≅ △CBD. If not, tell what else you need to know.

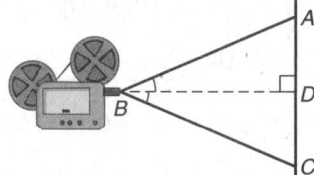

1. Hypotenuse-Leg

   No; you need to know that $\overline{AB} \cong \overline{CB}$.

2. Angle-Side-Angle

   Yes

3. Angle-Angle-Side

   Yes, if you use Third ∠s Thm. first.

Write which postulate, if any, can be used to prove the pair of triangles congruent.

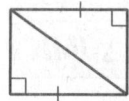

4. ____HL____

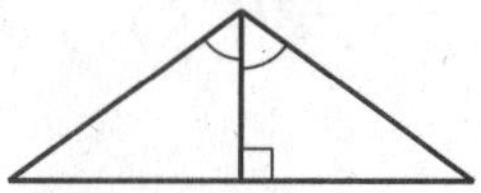

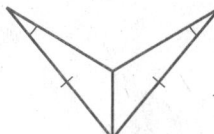

6. ____none____

5. ____ASA or AAS____

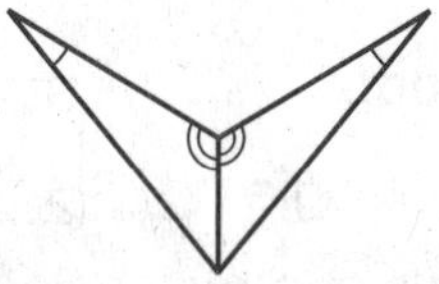

7. ____AAS or ASA____

Write a paragraph proof.

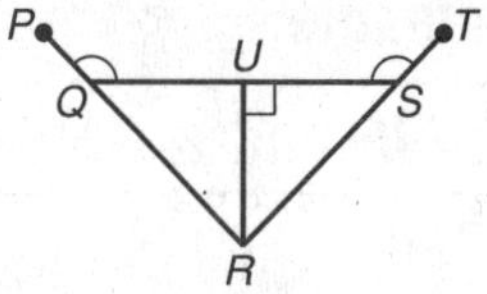

8. Given: ∠PQU ≅ ∠TSU,
   ∠QUR and ∠SUR are right angles.

   Prove: △RUQ ≅ △RUS

Possible answer: All right angles are congruent, so ∠QUR ≅ ∠SUR. ∠RQU and ∠PQU are supplementary and ∠RSU and ∠TSU are supplementary by the Linear Pair Theorem. But it is given that ∠PQU ≅ ∠TSU, so by the Congruent Supplements Theorem, ∠RQU ≅ ∠RSU. $\overline{RU} \cong \overline{RU}$ by the Reflexive Property of ≅, so △RUQ ≅ △RUS by AAS.

Name _____ Date _____ Class _____

LESSON 4-6 Practice B
Triangle Congruence: CPCTC

1. Heike Dreschler set the Woman's World Junior Record for the long jump in 1983. She jumped about 23.4 feet. The diagram shows two triangles and a pond. Explain whether Heike could have jumped the pond along path BA or along path CA. **Possible answer: Because ∠DCE ≅ ∠BCA by the**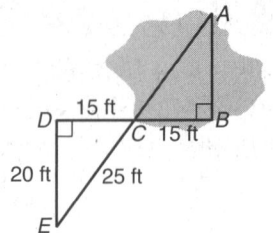

Vertical ∠s Thm. the triangles are congruent by ASA, and each side in △ABC has the

same length as its corresponding side in △EDC. Heike could jump about 23 ft. The

distance along path BA is 20 ft because BA corresponds with DE, so Heike could

have jumped this distance. The distance along path CA is 25 ft because CA

corresponds with CE, so Heike could not have jumped this distance.

Write a flowchart proof.

2. **Given:** ∠L ≅ ∠J, $\overline{KJ} \parallel \overline{LM}$
   Prove: ∠LKM ≅ ∠JMK

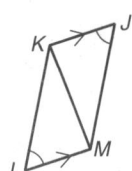

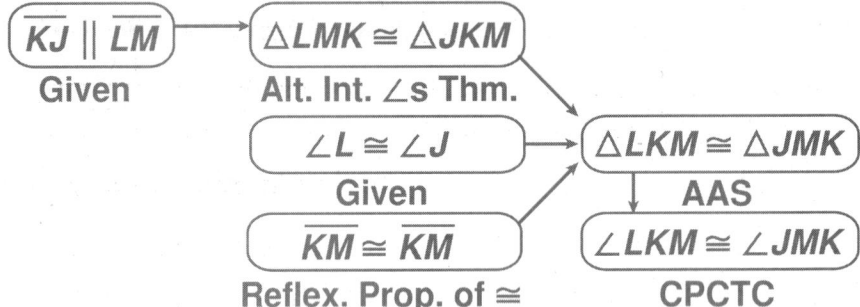

Write a two-column proof.

3. **Given:** FGHI is a rectangle.
   Prove: The diagonals of a rectangle have equal lengths. **Possible answer:**

Statements	Reasons
1. FGHI is a rectangle.	1. Given
2. $\overline{FI} \cong \overline{GH}$, ∠FIH and ∠GHI are right angles.	2. Def. of rectangle
3. ∠FIH ≅ ∠GHI	3. Rt. ∠ ≅ Thm.
4. $\overline{IH} \cong \overline{IH}$	4. Reflex. Prop. of ≅
5. △FIH ≅ △GHI	5. SAS
6. $\overline{FH} \cong \overline{GI}$	6. CPCTC
7. FH = GI	7. Def. of ≅ segs.

Name _____ Date _____ Class _____

LESSON 4-7 Practice B
Introduction to Coordinate Proof

Position an isosceles triangle with sides of 8 units, 5 units, and 5 units in the coordinate plane. Label the coordinates of each vertex. (*Hint:* Use the Pythagorean Theorem.)

1. Center the long side on the *x*-axis at the origin.

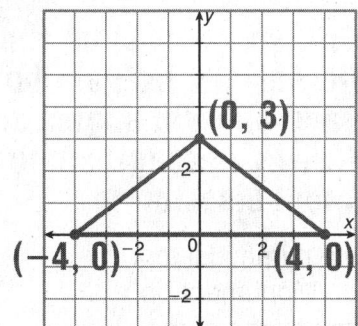

2. Place the long side on the *y*-axis centered at the origin.

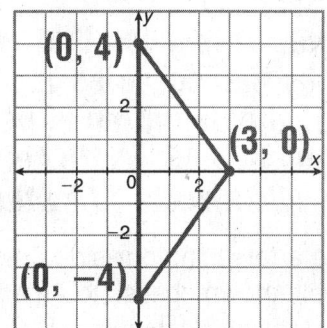

Write a coordinate proof.

3. **Given:** Rectangle *ABCD* has vertices *A*(0, 4), *B*(6, 4), *C*(6, 0), and *D*(0, 0). *E* is the midpoint of $\overline{DC}$. *F* is the midpoint of $\overline{DA}$.

   Prove: The area of rectangle *DEGF* is one-fourth the area of rectangle *ABCD*.

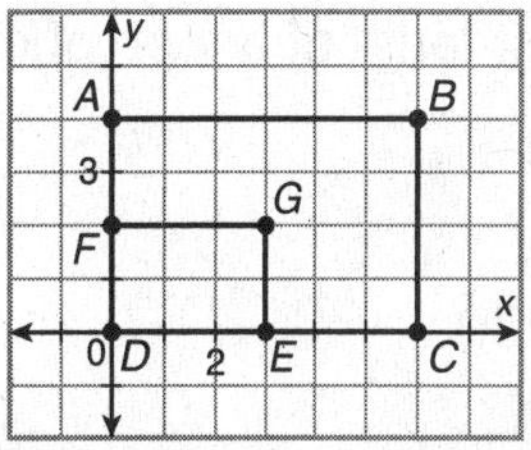

Possible answer: *ABCD* is a rectangle with width *AD* and length *DC*. The area of *ABCD* is (*AD*)(*DC*) or (4)(6) = 24 square units. By the Midpoint Formula, the coordinates of *E* are $\left(\frac{0+6}{2}, \frac{0+0}{2}\right) = (3, 0)$ and the coordinates of *F* are $\left(\frac{0+0}{2}, \frac{0+4}{2}\right) = (0, 2)$. The *x*-coordinate of *E* is the length of rectangle *DEGF*, and the *y*-coordinate of *F* is the width. So the area of *DEGF* is (3)(2) = 6 square units. Since $6 = \frac{1}{4}(24)$, the area of rectangle *DEGF* is one-fourth the area of rectangle *ABCD*.

Name _____ Date _____ Class _____

LESSON 4-8 Practice B
Isosceles and Equilateral Triangles

An altitude of a triangle is a perpendicular segment from a vertex to the line containing the opposite side. Write a paragraph proof that the altitude to the base of an isosceles triangle bisects the base.

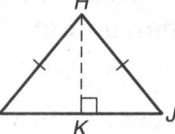

1. Given: $\overline{HI} \cong \overline{HJ}$, $\overline{HK} \perp \overline{IJ}$

   Prove: $\overline{HK}$ bisects $\overline{IJ}$.

Possible answer: It is given that $\overline{HI}$ is congruent to $\overline{HJ}$, so $\angle I$ must be congruent to $\angle J$ by the Isosceles Triangle Theorem. $\angle IKH$ and $\angle JKH$ are both right angles by the definition of perpendicular lines, and all right angles are congruent. Thus by AAS, $\triangle HKI$ is congruent to $\triangle HKJ$. $\overline{IK}$ is congruent to $\overline{KJ}$ by CPCTC, so HK bisects $\overline{IJ}$ by the definition of segment bisector.

2. An *obelisk* is a tall, thin, four-sided monument that tapers to a pyramidal top. The most well-known obelisk to Americans is the Washington Monument on the National Mall in Washington, D.C. Each face of the pyramidal top of the Washington Monument is an isosceles triangle. The height of each triangle is 55.5 feet, and the base of each triangle measures 34.4 feet. Find the length, to the nearest tenth of a foot, of one of the two equal legs of the triangle. __58.1 ft__

Find each value.

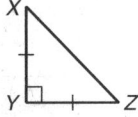

3. $m\angle X =$ __45°__

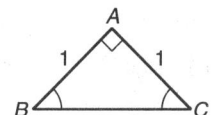

4. $BC =$ __$\sqrt{2}$__

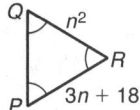

5. $PQ =$ __36 or 9__

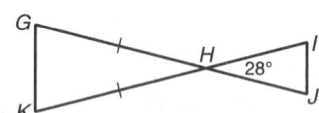

6. $m\angle K =$ __76°__

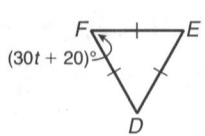

7. $t =$ __$\dfrac{4}{3}$__

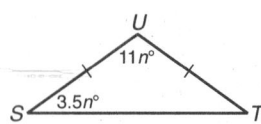

8. $n =$ __10__

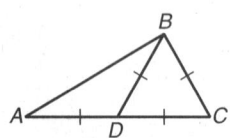

9. $m\angle A =$ __30°__

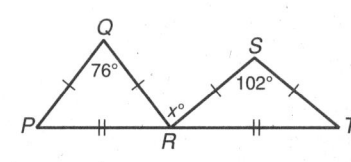

10. $x =$ __89__

Copyright © by Holt, Rinehart and Winston.
All rights reserved.

Holt Geometry

Name _____ Date _____ Class _____

LESSON 5-1 Practice B
Perpendicular and Angle Bisectors

Diana is in an archery competition. She stands at *A*, and the target is at *D*. Her competitors stand at *B* and *C*.

1. The distance from each of her competitors to her target is equal. Explain whether the flight path of Diana's arrow, $\overline{AD}$, must be a perpendicular bisector of $\overline{BC}$.

   Possible answer: The flight path of Diana's arrow does not have to be a perpendicular bisector of $\overline{BC}$. For that to be true, Diana must be equidistant from each of her competitors.

Use the figure for Exercises 2–5.

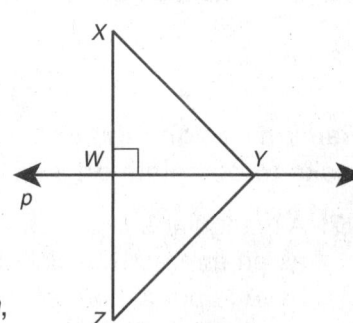

2. Given that line *p* is the perpendicular bisector of $\overline{XZ}$ and *XY* = 15.5, find *ZY*. _____**15.5**_____

3. Given that *XZ* = 38, *YX* = 27, and *YZ* = 27, find *ZW*. _____**19**_____

4. Given that line *p* is the perpendicular bisector of $\overline{XZ}$; *XY* = 4*n*, and *YZ* = 14, find *n*. _____$\frac{7}{2}$ **or 3.5**_____

5. Given that *XY* = *ZY*, *WX* = 6*x* − 1, and *XZ* = 10*x* + 16, find *ZW*. _____**53**_____

Use the figure for Exercises 6–9.

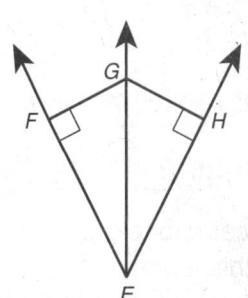

6. Given that *FG* = *HG* and m∠*FEH* = 55°, find m∠*GEH*. _____**27.5°**_____

7. Given that $\overrightarrow{EG}$ bisects ∠*FEH* and *GF* = $\sqrt{2}$, find *GH*. _____$\sqrt{2}$_____

8. Given that ∠*FEG* ≅ ∠*GEH*, *FG* = 10*z* − 30, and *HG* = 7*z* + 6, find *FG*. _____**90**_____

9. Given that *GF* = *GH*, m∠*GEF* = $\frac{8}{3}a°$, and m∠*GEH* = 24°, find *a*. _____**9**_____

Write an equation in point-slope form for the perpendicular bisector of the segment with the given endpoints.

10. *L*(4, 0), *M*(−2, 3)

    $y - \frac{3}{2} = 2(x - 1)$

11. *T*(0, −3), *U*(0, 1)

    $y + 1 = 0(x - 0)$ or $y + 1 = 0$

12. *A*(−1, 6), *B*(−3, −4)

    $y - 1 = -\frac{1}{5}(x + 2)$

Copyright © by Holt, Rinehart and Winston.
All rights reserved.

Holt Geometry

Name _____ Date _____ Class _____

LESSON 5-2 Practice B
Bisectors of Triangles

Use the figure for Exercises 1 and 2. $\overline{SV}$, $\overline{TV}$, and $\overline{UV}$ are perpendicular bisectors of the sides of $\triangle PQR$. Find each length.

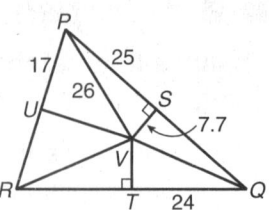

1. RV ___**26**___
2. TR ___**24**___

Find the circumcenter of the triangle with the given vertices.

3. A(0, 0), B(0, 5), C(5, 0)
   (___**2.5**___ , ___**2.5**___)

4. D(0, 7), E(–3, 1), F(3, 1)
   (___**0**___ , ___**3.25**___)

Use the figure for Exercises 7 and 8. $\overline{GJ}$ and $\overline{IJ}$ are angle bisectors of $\triangle GHI$. Find each measure.

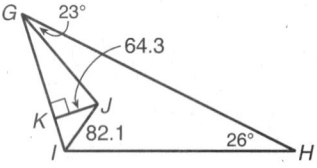

5. the distance from J to $\overline{GH}$ ___**64.3**___

6. $m\angle JGK$ ___**23°**___

Raleigh designs the interiors of cars. He is given two tasks to complete on a new production model.

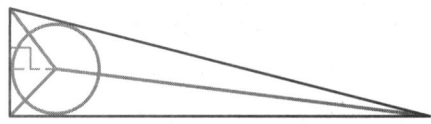

7. A triangular surface as shown in the figure is molded into the driver's side door as an armrest. Raleigh thinks he can fit a cup holder into the triangle, but he'll have to put the largest possible circle into the triangle. Explain how Raleigh can do this. Sketch his design on the figure.

   Possible answer: Raleigh needs to find the incircle of the triangle. The incircle just touches all three sides of the triangle, so it is the largest circle that will fit. The incenter can be found by drawing the angle bisector from each vertex of the triangle. The incircle is drawn with the incenter as the center and a radius equal to the distance to one of the sides.

8. The car's logo is the triangle shown in the figure. Raleigh has to use this logo as the center of the steering wheel. Explain how Raleigh can do this. Sketch his design on the figure.

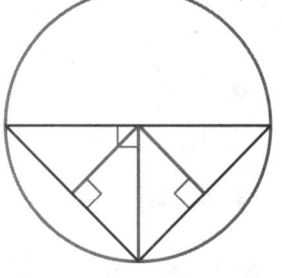

   Possible answer: Raleigh needs to find the circumcircle of the triangle. The circumcircle just touches all three vertices of the triangle, so it fits just around it. The circumcenter can be found by drawing the perpendicular bisectors of the sides of the triangle. The circumcircle is drawn with the circumcenter as center and a radius equal to the distance from the center to one of the vertices.

Name _____ Date _____ Class _____

LESSON 5-3 Practice B
Medians and Altitudes of Triangles

Use the figure for Exercises 1–4. $GB = 12\frac{2}{3}$ and $CD = 10$.
Find each length.

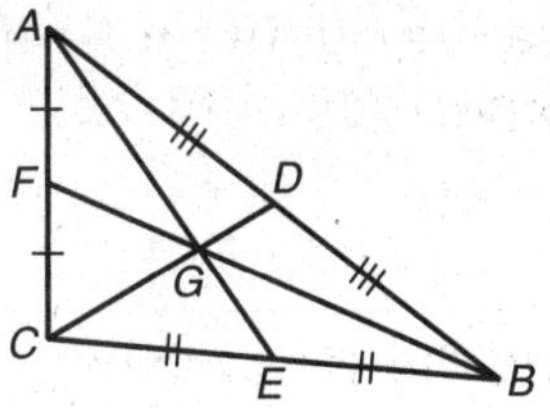

1. FG __$6\frac{1}{3}$__
2. BF __19__
3. GD __$3\frac{1}{3}$__
4. CG __$6\frac{2}{3}$__

5. A triangular compass needle will turn most easily if it is attached to the compass face through its centroid. Find the coordinates of the centroid.   (__1__ , __1.9__)

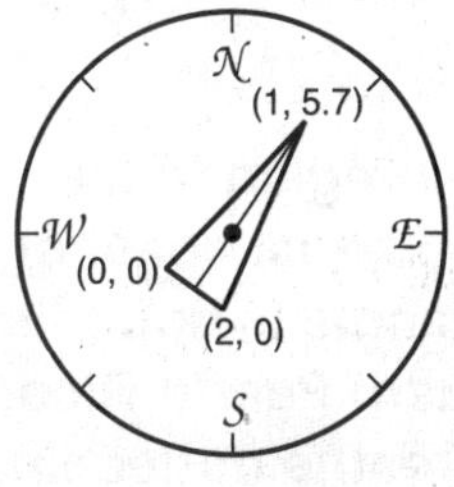

Find the orthocenter of the triangle with the given vertices.

6. $X(-5, 4)$, $Y(2, -3)$, $Z(1, 4)$
   (__2__ , __5__)

7. $A(0, -1)$, $B(2, -3)$, $C(4, -1)$
   (__2__ , __-3__)

Use the figure for Exercises 8 and 9. $\overline{HL}$, $\overline{IM}$, and $\overline{JK}$ are medians of $\triangle HIJ$.

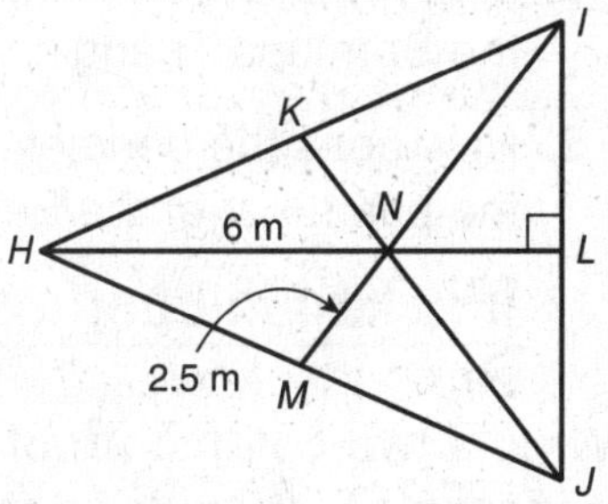

8. Find the area of the triangle.   __36 m^2__

9. If the perimeter of the triangle is 49 meters, then find the length of $\overline{MH}$. (Hint: What kind of a triangle is it?)
   __10.25 m__

10. Two medians of a triangle were cut apart at the centroid to make the four segments shown below. Use what you know about the Centroid Theorem to reconstruct the original triangle from the four segments shown. Measure the side lengths of your triangle to check that you constructed medians. (Note: There are many possible answers.)

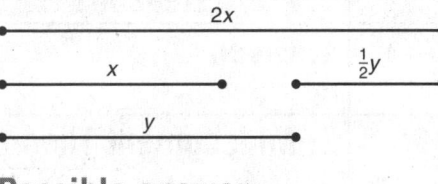

Possible answer:

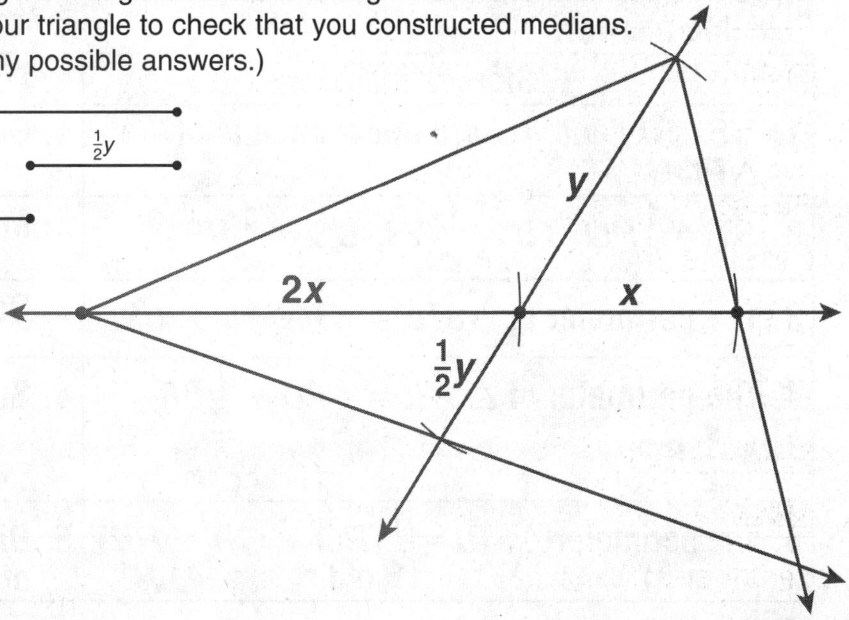

Name _____ Date _____ Class _____

LESSON 5-4 Practice B
The Triangle Midsegment Theorem

Use the figure for Exercises 1–6. Find each measure.

1. HI __9.1__   2. DF __35__
3. GE __9.1__   4. m∠HIF __58°__
5. m∠HGD __122°__   6. m∠D __58°__

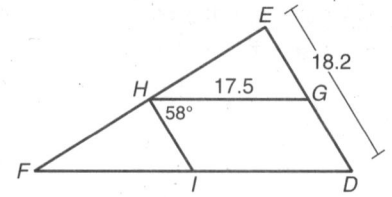

The Bermuda Triangle is a region in the Atlantic Ocean off the southeast coast of the United States. The triangle is bounded by Miami, Florida; San Juan, Puerto Rico; and Bermuda. In the figure, the dotted lines are midsegments.

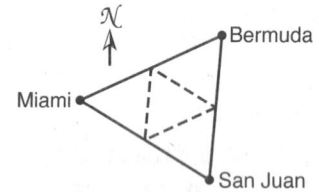

	Dist. (mi)
Miami to San Juan	1038
Miami to Bermuda	1042
Bermuda to San Juan	965

7. Use the distances in the chart to find the perimeter of the Bermuda Triangle.   __3045 mi__

8. Find the perimeter of the midsegment triangle within the Bermuda Triangle.   __1522.5 mi__

9. How does the perimeter of the midsegment triangle compare to the perimeter of the Bermuda Triangle?

__It is half the perimeter of the Bermuda Triangle.__

Write a two-column proof that the perimeter of a midsegment triangle is half the perimeter of the triangle.

10. **Given:** $\overline{US}$, $\overline{ST}$, and $\overline{TU}$ are midsegments of $\triangle PQR$.

Prove: The perimeter of $\triangle STU = \frac{1}{2}(PQ + QR + RP)$.

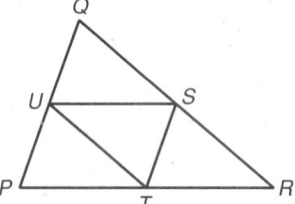

Possible answer:

Statements	Reasons
1. $\overline{US}$, $\overline{ST}$, and $\overline{TU}$ are midsegments of $\triangle PQR$.	1. Given
2. $ST = \frac{1}{2}PQ$, $TU = \frac{1}{2}QR$, $US = \frac{1}{2}RP$	2. Midsegment Theorem
3. The perimeter of $\triangle STU = ST + TU + US$.	3. Definition of perimeter
4. The perimeter of $\triangle STU = \frac{1}{2}PQ + \frac{1}{2}QR + \frac{1}{2}RP$.	4. Substitution
5. The perimeter $\triangle STU = \frac{1}{2}(PQ + QR + RP)$	5. Distributive Property of =

Name _____ Date _____ Class _____

Practice B
LESSON 5-5 Indirect Proof and Inequalities in One Triangle

Write an indirect proof that the angle measures of a triangle cannot add to more than 180°.

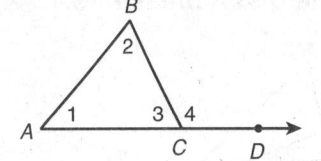

1. State the assumption that starts the indirect proof.

   $m\angle 1 + m\angle 2 + m\angle 3 > 180°$

2. Use the Exterior Angle Theorem and the Linear Pair Theorem to write the indirect proof.
   Possible answer: Assume that $m\angle 1 + m\angle 2 + m\angle 3 > 180°$. $\angle 4$ is an exterior angle of $\triangle ABC$, so by the Exterior Angle Theorem, $m\angle 1 + m\angle 2 = m\angle 4$. $\angle 3$ and $\angle 4$ are a linear pair, so by the Linear Pair Theorem, $m\angle 3 + m\angle 4 = 180°$. Substitution leads to the conclusion that $m\angle 1 + m\angle 2 + m\angle 3 = 180°$, which contradicts the assumption. Thus the assumption is false, and the sum of the angle measures of a triangle cannot add to more than 180°.

3. Write the angles of $\triangle DEF$ in order from smallest to largest.

   $\angle F; \angle D; \angle E$

4. Write the sides of $\triangle GHI$ in order from shortest to longest.

   $\overline{HI}; \overline{GH}; \overline{GI}$

Tell whether a triangle can have sides with the given lengths. If not, explain why not.

5. 8, 8, 16 **no; 8 + 8 = 16**
6. 0.5, 0.7, 0.3 **yes**
7. $10\frac{1}{2}$, 4, 14 **yes**
8. $3x + 2$, x^2, $2x$ when $x = 4$ **yes**
9. $3x + 2$, x^2, $2x$ when $x = 6$ **no; 12 + 20 < 36**

The lengths of two sides of a triangle are given. Find the range of possible lengths for the third side.

10. 8.2 m, 3.5 m

    $4.7 \text{ m} < s < 11.7 \text{ m}$

11. 298 ft, 177 ft

    $121 \text{ ft} < s < 475 \text{ ft}$

12. $3\frac{1}{2}$ mi, 4 mi

    $\frac{1}{2} \text{ mi} < s < 7\frac{1}{2} \text{ mi}$

13. The annual Cheese Rolling happens in May at Gloucestershire, England. As the name suggests, large, 7–9 pound wheels of cheese are rolled down a steep hill, and people chase after them. The first person to the bottom wins cheese. Renaldo wants to go to the Cheese Rolling. He plans to leave from Atlanta and fly into London (4281 miles). On the return, he will fly back from London to New York City (3470 miles) to visit his aunt. Then Renaldo heads back to Atlanta. Atlanta, New York City, and London do not lie on the same line. Find the range of the total distance Renaldo could travel on his trip.

    Renaldo could travel between 8562 miles and 15,502 miles.

LESSON 5-6 Practice B
Inequalities in Two Triangles

Compare the given measures.

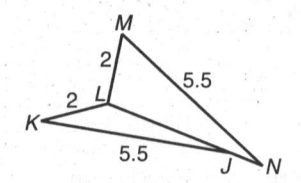

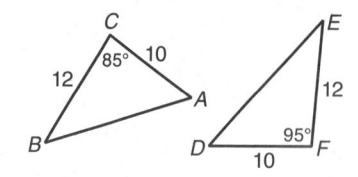

1. m∠K and m∠M

   m∠K < m∠M

2. AB and DE

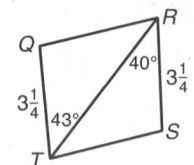

   AB < DE

3. QR and ST

   QR > ST

Find the range of values for x.

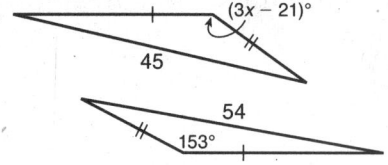

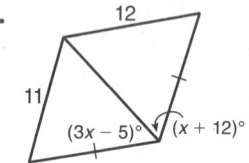

4. 7 < x < 58

5. −2 < x < 10.5

6. $\frac{5}{3} < x < \frac{17}{2}$

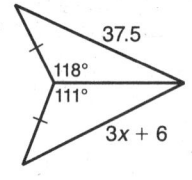

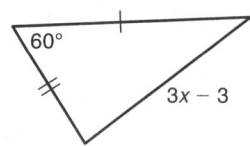

7. x > 4

8. You have used a compass to copy and bisect segments and angles and to draw arcs and circles. A compass has a drawing leg, a pivot leg, and a hinge at the angle between the legs. Explain why and how the measure of the angle at the hinge changes if you draw two circles with different diameters.

   Possible answer: The legs of a compass and the length spanned by it form a triangle, but the lengths of the legs cannot change. Therefore any two settings of the compass are subject to the Hinge Theorem. To draw a larger-diameter circle, the measure of the hinge angle must be made larger. To draw a smaller-diameter circle, the measure of the hinge angle must be made smaller.

Name _____ Date _____ Class _____

LESSON 5-7 Practice B
The Pythagorean Theorem

Find the value of x. Give your answer in simplest radical form.

1. [triangle with sides x, 6, 5]

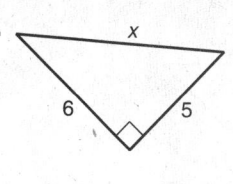

$\sqrt{61}$

2. [triangle with sides 13, x, 15]

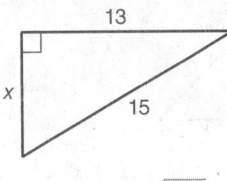

$2\sqrt{14}$

3. [triangle with sides 14, x+2, x]

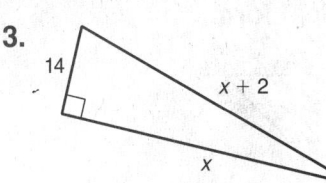

48

4. The aspect ratio of a TV screen is the ratio of the width to the height of the image. A regular TV has an aspect ratio of 4 : 3. Find the height and width of a 42-inch TV screen to the nearest tenth of an inch. (The measure given is the length of the diagonal across the screen.)

height: 25.2 in.; width: 33.6 in.

5. A "wide-screen" TV has an aspect ratio of 16 : 9. Find the length of a diagonal on a wide-screen TV screen that has the same height as the screen in Exercise 4.

51.4 in.

Find the missing side lengths. Give your answer in simplest radical form. Tell whether the side lengths form a Pythagorean Triple.

6. [triangle with sides 6, 6.5]

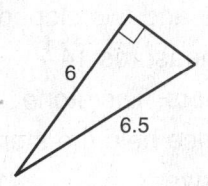

2.5; no

7. [triangle with sides 20, 15]

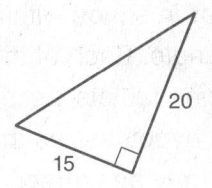

25; yes

8. [triangle with sides 9, 3]

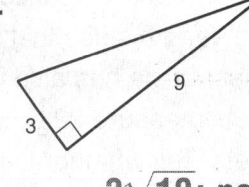

$3\sqrt{10}$; no

Tell whether the measures can be the side lengths of a triangle. If so, classify the triangle as acute, obtuse, or right.

9. 15, 18, 20

yes; acute

10. 7, 8, 11

yes; obtuse

11. 6, 7, $3\sqrt{13}$

yes; obtuse

12. Kitty has a triangle with sides that measure 16, 8, and 13. She does some calculations and finds that 256 + 64 > 169. Kitty concludes that the triangle is obtuse. Evaluate Kitty's conclusion and Kitty's reasoning.

Possible answer: The triangle is obtuse, so Kitty is correct. But Kitty did not use the Pythagorean Inequalities Theorem correctly. The measure of the longest side should be substituted for c, so 169 + 64 < 256 is the inequality that shows that the triangle is obtuse.

Practice B
5-8 Applying Special Right Triangles

Find the value of x in each figure. Give your answer in simplest radical form.

1. 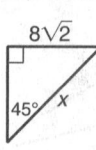 __16__

2. 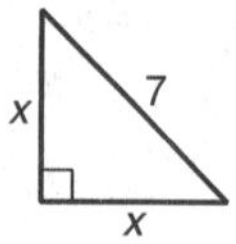 __$\frac{7\sqrt{2}}{2}$__

3. 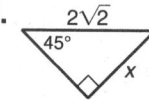 __2__

Find the values of x and y. Give your answers in simplest radical form.

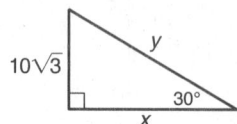

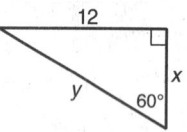

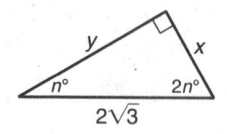

4. x = __30__ y = __$20\sqrt{3}$__

5. x = __$4\sqrt{3}$__ y = __$8\sqrt{3}$__

6. x = __$\sqrt{3}$__ y = __3__

Lucia is an archaeologist trekking through the jungle of the Yucatan Peninsula. She stumbles upon a stone structure covered with creeper vines and ferns. She immediately begins taking measurements of her discovery. (Hint: Drawing some figures may help.)

7. Around the perimeter of the building, Lucia finds small alcoves at regular intervals carved into the stone. The alcoves are triangular in shape with a horizontal base and two sloped equal-length sides that meet at a right angle. Each of the sloped sides measures $14\frac{1}{4}$ inches. Lucia has also found several stone tablets inscribed with characters. The stone tablets measure $22\frac{1}{8}$ inches long. Lucia hypothesizes that the alcoves once held the stone tablets. Tell whether Lucia's hypothesis may be correct. Explain your answer.

Possible answer: Lucia's hypothesis cannot be correct. The base of the alcove is $\frac{57\sqrt{2}}{4}$ inches or just over 20 inches long, so a $22\frac{1}{8}$-inch tablet could not fit.

8. Lucia also finds several statues around the building. The statues measure $9\frac{7}{16}$ inches tall. She wonders whether the statues might have been placed in the alcoves. Tell whether this is possible. Explain your answer.

Possible answer: To find the height of a 45°-45°-90° triangle, draw a perpendicular to the hypotenuse. This makes another smaller 45°-45°-90° triangle whose hypotenuse is the length of one of the legs of the larger triangle. The height of the alcove is $\frac{57\sqrt{2}}{8}$ inches or about 10 inches, so the statues could have been placed in the alcoves.

Name _____ Date _____ Class _____

LESSON 6-1 Practice B
Properties and Attributes of Polygons

Tell whether each figure is a polygon. If it is a polygon, name it by the number of its sides.

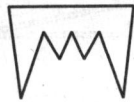

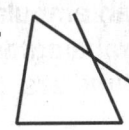

1. polygon; nonagon

2. not a polygon

3. not a polygon

4. For a polygon to be regular, it must be both equiangular and equilateral. Name the only type of polygon that must be regular if it is equiangular. __triangle__

Tell whether each polygon is regular or irregular. Then tell whether it is concave or convex.

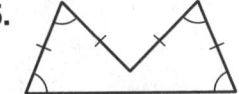

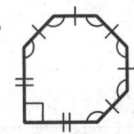

5. irregular; concave

6. regular; convex

7. irregular; convex

8. Find the sum of the interior angle measures of a 14-gon. __2160°__

9. Find the measure of each interior angle of hexagon ABCDEF.
m∠A = 60°; m∠B = m∠D = m∠F = 150°;
m∠C = 120°; m∠E = 90°

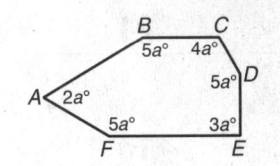

10. Find the value of n in pentagon PQRST. __24__

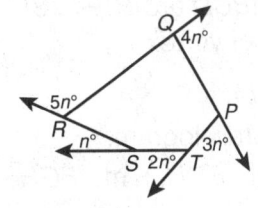

Before electric or steam power, a common way to power machinery was with a waterwheel. The simplest form of waterwheel is a series of paddles on a frame partially submerged in a stream. The current in the stream pushes the paddles forward and turns the frame. The power of the turning frame can then be used to drive machinery to saw wood or grind grain. The waterwheel shown has a frame in the shape of a regular octagon.

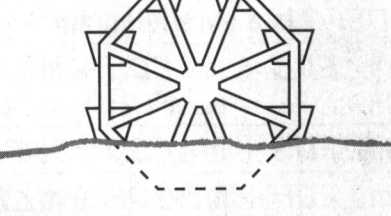

11. Find the measure of one interior angle of the waterwheel. __135°__

12. Find the measure of one exterior angle of the waterwheel. __45°__

Name _____ Date _____ Class _____

LESSON 6-2 Practice B
Properties of Parallelograms

A gurney is a wheeled cot or stretcher used in hospitals. Many gurneys are made so that the base will fold up for easy storage in an ambulance. When partially folded, the base forms a parallelogram. In □STUV, VU = 91 centimeters, UW = 108.8 centimeters, and m∠TSV = 57°. Find each measure.

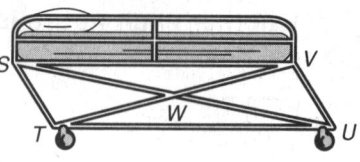

1. SW   2. TS   3. US
   108.8 cm   91 cm   217.6 cm

4. m∠SVU   5. m∠STU   6. m∠TUV
   123°   123°   57°

JKLM is a parallelogram. Find each measure.

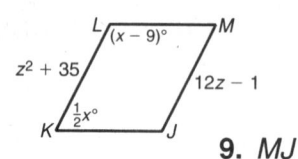

7. m∠L   8. m∠K   9. MJ
   117°   63°   71

VWXY is a parallelogram. Find each measure.

10. VX   11. XZ
    21   10.5

12. ZW   13. WY
    15   30

14. Three vertices of □ABCD are B(−3, 3), C(2, 7), and D(5, 1). Find the coordinates of vertex A.   (0, −3)

Write a two-column proof.

15. Given: DEFG is a parallelogram.
    Prove: m∠DHG = m∠EDH + m∠FGH   **Possible answer:**

Statements	Reasons
1. DEFG is a parallelogram.	1. Given
2. m∠EDG = m∠EDH + m∠GDH, m∠FGD = m∠FGH + m∠DGH	2. Angle Add. Post.
3. m∠EDG + m∠FGD = 180°	3. □ → cons. ∠s supp.
4. m∠EDH + m∠GDH + m∠FGH + m∠DGH = 180°	4. Subst. (Steps 2, 3)
5. m∠GDH + m∠DGH + m∠DHG = 180°	5. Triangle Sum Thm.
6. m∠GDH + m∠DGH + m∠DHG = m∠EDH + m∠GDH + m∠FGH + m∠DGH	6. Trans. Prop. of =
7. m∠DHG = m∠EDH + m∠FGH	7. Subtr. Prop. of =

Holt Geometry

LESSON 6-3
Practice B
Conditions for Parallelograms

For Exercises 1 and 2, determine whether the figure is a parallelogram for the given values of the variables. Explain your answers.

1. $x = 9$ and $y = 11$

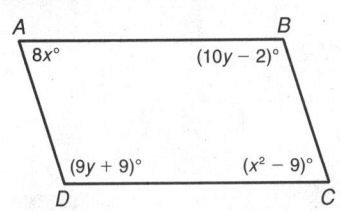

2. $a = 4.3$ and $b = 13$

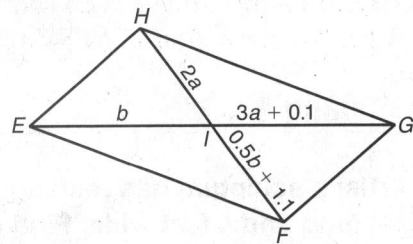

ABCD is a parallelogram. $m\angle A$ = $m\angle C$ = 72° and $m\angle B$ = $m\angle D$ = 108°

EFGH is not a parallelogram. HI = 8.6 and FI = 7.6. $\overline{EG}$ does not bisect $\overline{HF}$.

Determine whether each quadrilateral must be a parallelogram. Justify your answers.

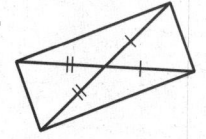

3. No, the diagonals do not necessarily bisect each other.

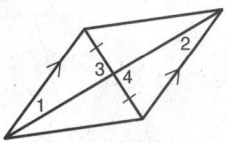

4. Yes, the triangles with numbered angles are ≅ by AAS. By CPCTC, the parallel sides are congruent.

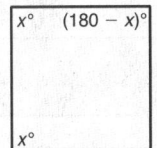

5. No, $x°$ + $x°$ may not be 180°.

Use the given method to determine whether the quadrilateral with the given vertices is a parallelogram.

6. Find the slopes of all four sides: $J(-4, -1)$, $K(-7, -4)$, $L(2, -10)$, $M(5, -7)$
   slope of $\overline{JK}$ = slope of $\overline{LM}$ = 1; slope of $\overline{KL}$ = slope of $\overline{JM}$ = $-\frac{2}{3}$; JKLM is a parallelogram.

7. Find the lengths of all four sides: $P(2, 2)$, $Q(1, -3)$, $R(-4, 2)$, $S(-3, 7)$
   $PQ = RS = \sqrt{26}$; $QR = PS = 5\sqrt{2}$; PQRS is a parallelogram.

8. Find the slopes and lengths of one pair of opposite sides:
   $T\left(\frac{3}{2}, -2\right)$, $U\left(\frac{3}{2}, 4\right)$, $V\left(-\frac{1}{2}, 0\right)$, $W\left(-\frac{1}{2}, -6\right)$ Possible answer: $UV = TW = 2\sqrt{5}$; slope of $\overline{UV}$ = slope of $\overline{TW}$ = 2; TUVW is a parallelogram.

Name _____ Date _____ Class _____

LESSON 6-4 Practice B
Properties of Special Parallelograms

Tell whether each figure must be a rectangle, rhombus, or square based on the information given. Use the most specific name possible.

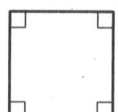

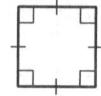

1. **rectangle**

2. **square**

3. **rhombus**

A modern artist's sculpture has rectangular faces. The face shown here is 9 feet long and 4 feet wide. Find each measure in simplest radical form. (*Hint:* Use the Pythagorean Theorem.)

4. DC = __9 feet__

5. AD = __4 ft__

6. DB = __$\sqrt{97}$ feet__

7. AE = __$\dfrac{\sqrt{97}}{2}$ ft__

VWXY is a rhombus. Find each measure.

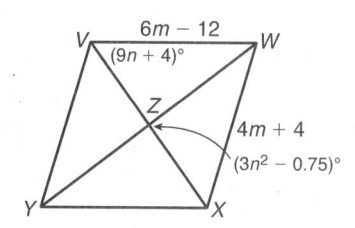

8. XY = __36__

9. m∠YVW = __107°__

10. m∠VYX = __73°__

11. m∠XYZ = __36.5°__

12. The vertices of square *JKLM* are *J*(−2, 4), *K*(−3, −1), *L*(2, −2), and *M*(3, 3). Find each of the following to show that the diagonals of square *JKLM* are congruent perpendicular bisectors of each other.

JL = __$2\sqrt{13}$__   KM = __$2\sqrt{13}$__

slope of $\overline{JL}$ = __$-\dfrac{3}{2}$__   slope of $\overline{KM}$ = __$\dfrac{2}{3}$__

midpoint of $\overline{JL}$ = (__0__ , __1__)   midpoint of $\overline{KM}$ = (__0__ , __1__)

Write a paragraph proof.

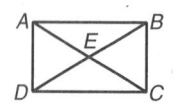

13. **Given:** *ABCD* is a rectangle.
    Prove: ∠EDC ≅ ∠ECD

Possible answer: *ABCD* is a rectangle, so $\overline{AC}$ is congruent to $\overline{BD}$. Because *ABCD* is a rectangle, it is also a parallelogram. Because *ABCD* is a parallelogram, its diagonals bisect each other. By the definition of bisector, EC = ½AC and ED = ½BD. But by the definition of congruent segments, AC = BD. So substitution and the Transitive Property of Equality show that EC = ED. Because $\overline{EC} \cong \overline{ED}$, △ECD is an isosceles triangle. The base angles of an isosceles triangle are congruent, so ∠EDC ≅ ∠ECD.

Name _____ Date _____ Class _____

LESSON 6-5 Practice B
Conditions for Special Parallelograms

1. On the National Mall in Washington, D.C., a reflecting pool lies between the Lincoln Memorial and the World War II Memorial. The pool has two 2300-foot-long sides and two 150-foot-long sides. Tell what additional information you need to know in order to determine whether the reflecting pool is a rectangle. (*Hint:* Remember that you have to show it is a parallelogram first.)

   Possible answer: To know that the reflecting pool is a parallelogram, the congruent sides must be opposite each other. If this is true, then knowing that one angle in the pool is a right angle or that the diagonals are congruent proves that the pool is a rectangle.

Use the figure for Exercises 2–5. Determine whether each conclusion is valid. If not, tell what additional information is needed to make it valid.

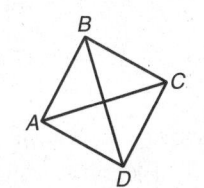

2. Given: $\overline{AC}$ and $\overline{BD}$ bisect each other. $\overline{AC} \cong \overline{BD}$
   Conclusion: ABCD is a square.
   Not valid; possible answer: you need to know that $\overline{AC} \perp \overline{BD}$.

3. Given: $\overline{AC} \perp \overline{BD}$, $\overline{AB} \cong \overline{BC}$
   Conclusion: ABCD is a rhombus. **Not valid;**
   possible answer: you need to know that $\overline{AC}$ and $\overline{BD}$ bisect each other.

4. Given: $\overline{AB} \cong \overline{DC}$, $\overline{AD} \cong \overline{BC}$, $m\angle ADB = m\angle ABD = 45°$
   Conclusion: ABCD is a square.
   valid

5. Given: $\overline{AB} \parallel \overline{DC}$, $\overline{AD} \cong \overline{BC}$, $\overline{AC} \cong \overline{BD}$
   Conclusion: ABCD is a rectangle.
   Not valid; possible answer: you need to know that $\overline{AD} \parallel \overline{BC}$.

Find the lengths and slopes of the diagonals to determine whether a parallelogram with the given vertices is a rectangle, rhombus, or square. Give all names that apply.

6. E(−2, −4), F(0, −1), G(−3, 1), H(−5, −2)   **rectangle, rhombus, square**
   EG = **√26**   FH = **√26**
   slope of $\overline{EG}$ = **−5**   slope of $\overline{FH}$ = **1/5**

7. P(−1, 3), Q(−2, 5), R(0, 4), S(1, 2)   **rhombus**
   PR = **√2**   QS = **3√2**
   slope of $\overline{PR}$ = **1**   slope of $\overline{QS}$ = **−1**

Name _____ Date _____ Class _____

LESSON 6-6 Practice B
Properties of Kites and Trapezoids

In kite ABCD, m∠BAC = 35° and m∠BCD = 44°.
For Exercises 1–3, find each measure.

1. m∠ABD   2. m∠DCA   3. m∠ABC
   55°   **22°**   **123°**

4. Find the area of △EFG.   **60 unit²**

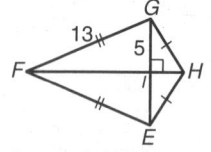

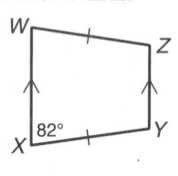

5. Find m∠Z.
   98°

6. KM = 7.5, and NM = 2.6. Find LN.
   4.9

7. Find the value of n so that PQRS is isosceles.
   n = 11.5

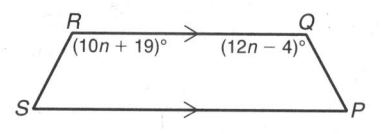

8. Find the value of x so that EFGH is isosceles.
   x = 12 or −12

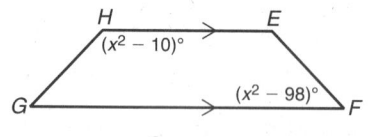

9. BD = 7a − 0.5, and AC = 5a + 2.3. Find the value of a so that ABCD is isosceles.
   a = 1.4

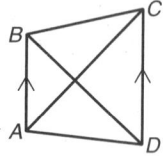

10. QS = 8z², and RT = 6z² + 38. Find the value of z so that QRST is isosceles.
    z = √19 or −√19

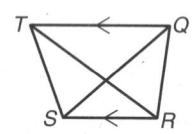

Use the figure for Exercises 11 and 12. The figure shows a *ziggurat*. A ziggurat is a stepped, flat-topped pyramid that was used as a temple by ancient peoples of Mesopotamia. The dashed lines show that a ziggurat has sides roughly in the shape of a trapezoid.

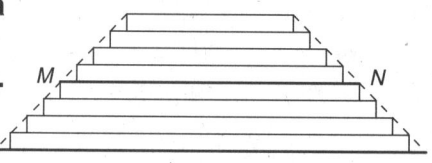

11. Each "step" in the ziggurat has equal height. Give the vocabulary term for $\overline{MN}$.
    trapezoid midsegment

12. The bottom of the ziggurat is 27.3 meters long, and the top of the ziggurat is 11.6 meters long. Find MN.
    19.45 m

Name _____ Date _____ Class _____

LESSON 7-1 Practice B
Ratio and Proportion

Use the graph for Exercises 1–3. Write a ratio expressing the slope of each line.

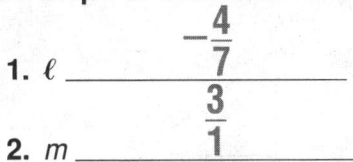

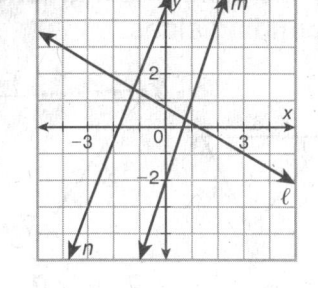

1. ℓ _____ $-\dfrac{4}{7}$_____

2. m _____ $\dfrac{3}{1}$_____

3. n _____ $\dfrac{5}{2}$_____

4. The ratio of the angle measures in a quadrilateral is 1 : 4 : 5 : 6. Find each angle measure. _____ 22.5°; 90°; 112.5°; 135° _____

5. The ratio of the side lengths in a rectangle is 5 : 2 : 5 : 2, and its area is 90 square feet. Find the side lengths. _____ 15 ft; 6 ft _____

For part of her homework, Celia measured the angles and the lengths of the sides of two triangles. She wrote down the ratios for angle measures and side lengths. Two of the ratios were 4 : 7 : 8 and 3 : 8 : 13.

6. When Celia got to school the next day, she couldn't remember which ratio was for angles and which was for sides. Tell which must be the ratio of the lengths of the sides. Explain your answer.

   4 : 7 : 8 must be the ratio of the lengths of the sides. The Triangle Inequality Theorem states that no side of a triangle can be longer than the sum of the lengths of the other two sides. If the ratio of the side lengths was 3 : 8 : 13, one side would be longer than the sum of the other two sides.

7. Find the measures of the angles of one of Celia's triangles. _____ 22.5°; 60°; 97.5° _____

Solve each proportion.

8. $\dfrac{28}{p} = \dfrac{42}{3}$

   $p =$ _____ 2 _____

9. $\dfrac{28}{24} = \dfrac{q}{102}$

   $q =$ _____ 119 _____

10. $\dfrac{3}{4.5} = \dfrac{7}{r}$

    $r =$ _____ 10.5 _____

11. $\dfrac{9}{s} = \dfrac{s}{25}$

    $s =$ _____ ±15 _____

12. $\dfrac{50}{2t+4} = \dfrac{2t+4}{2}$

    $t =$ _____ 3, −7 _____

13. $\dfrac{u+3}{8} = \dfrac{5}{u-3}$

    $u =$ _____ ±7 _____

14. Given that $12a = 20b$, find the ratio of a to b in simplest form. _____ 5 to 3 _____

15. Given that $34x = 51y$, find the ratio $x : y$ in simplest form. _____ 3 : 2 _____

Name _____ Date _____ Class _____

LESSON 7-2 Practice B
Ratios in Similar Polygons

Identify the pairs of congruent corresponding angles and the corresponding sides.

1.

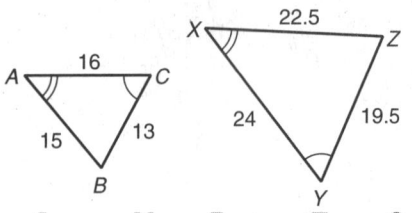

$\angle A \cong \angle X; \angle B \cong \angle Z; \angle C \cong \angle Y;$
$\dfrac{AC}{XY} = \dfrac{AB}{XZ} = \dfrac{BC}{ZY} = \dfrac{2}{3}$

2.

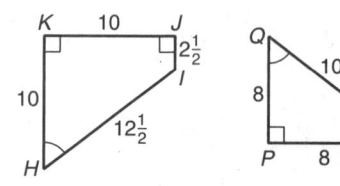

$\angle H \cong \angle Q; \angle I \cong \angle R; \angle J \cong \angle S;$
$\angle K \cong \angle P;$
$\dfrac{KJ}{PS} = \dfrac{KH}{PQ} = \dfrac{HI}{QR} = \dfrac{JI}{SR} = \dfrac{5}{4}$

Determine whether the polygons are similar. If so, write the similarity ratio and a similarity statement. If not, explain why not.

3. parallelograms EFGH and TUVW

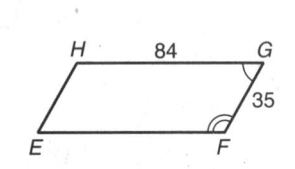

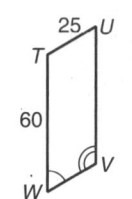

yes; $\dfrac{7}{5}$; Possible answer:

▱EFGH ~ ▱WTUV

4. △CDE and △LMN

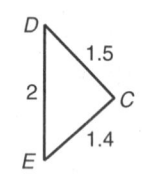

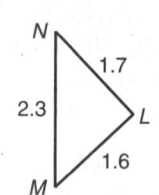

No; sides cannot be matched to have corresponding sides proportional.

Tell whether the polygons must be similar based on the information given in the figures.

5. 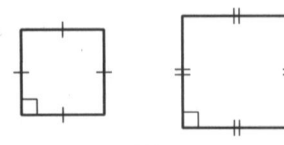 yes

6. 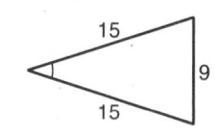 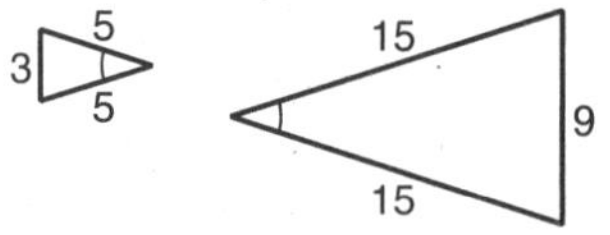 yes

7. 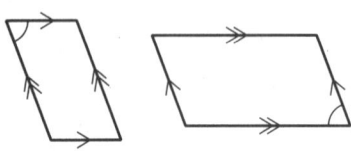 no

8.  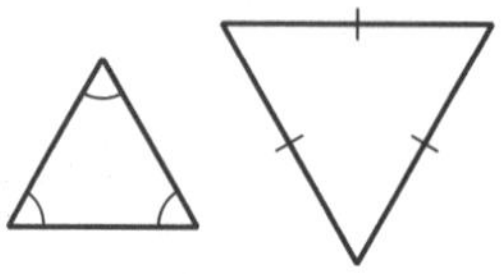 yes

44   Holt Geometry

Name _____ Date _____ Class _____

LESSON 7-3 Practice B
Triangle Similarity: AA, SSS, SAS

For Exercises 1 and 2, explain why the triangles are similar and write a similarity statement.

1.

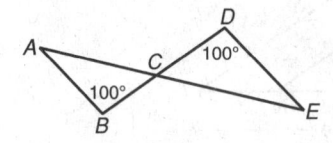

Possible answer: ∠ACB and ∠ECD are congruent vertical angles. m∠B = m∠D = 100°, so ∠B ≅ ∠D. Thus, △ABC ~ △EDC by AA ~.

2.

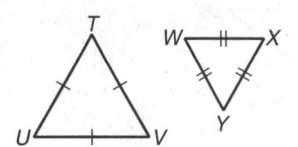

Possible answer: Every equilateral triangle is also equiangular, so each angle in both triangles measures 60°. Thus, △TUV ~ △WXY by AA ~.

For Exercises 3 and 4, verify that the triangles are similar. Explain why.

3. △JLK and △JMN

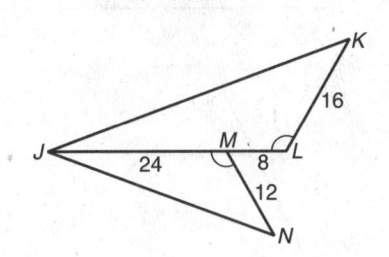

Possible answer: It is given that ∠JMN ≅ ∠L. $\frac{KL}{MN} = \frac{JL}{JM} = \frac{4}{3}$.

Thus, △JLK ~ △JMN by SAS ~.

4. △PQR and △UTS

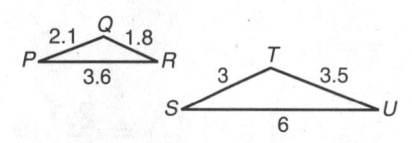

Possible answer: $\frac{PQ}{UT} = \frac{QR}{TS} = \frac{PR}{US} = \frac{3}{5}$.

Thus, △PQR ~ △UTS by SSS ~.

For Exercise 5, explain why the triangles are similar and find the stated length.

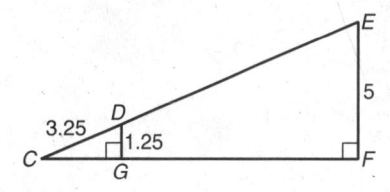

5. DE

Possible answer: ∠C ≅ ∠C by the Reflexive Property. ∠CGD and ∠F are right angles, so they are congruent. Thus, △CDG ~ △CEF by AA ~.

DE = 9.75

LESSON 7-4 Practice B
Applying Properties of Similar Triangles

Find each length.

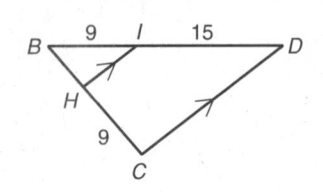

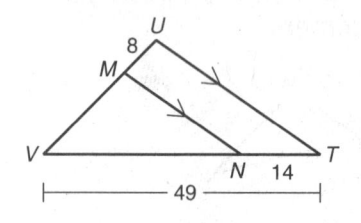

1. BH ___5.4___

2. MV ___20___

Verify that the given segments are parallel.

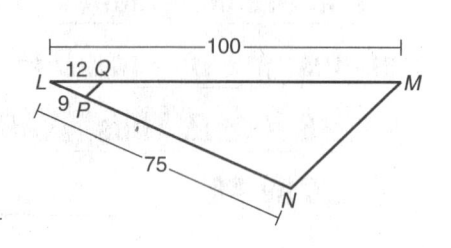

3. $\overline{PQ}$ and $\overline{NM}$

PN = 66 and QM = 88. $\dfrac{LP}{PN} = \dfrac{9}{66} = \dfrac{3}{22}$ and $\dfrac{LQ}{QM} = \dfrac{12}{88} = \dfrac{3}{22}$. Because $\dfrac{LP}{PN} = \dfrac{LQ}{QM}$, $\overline{PQ} \parallel \overline{NM}$ by the Conv. of the △ Proportionality Thm.

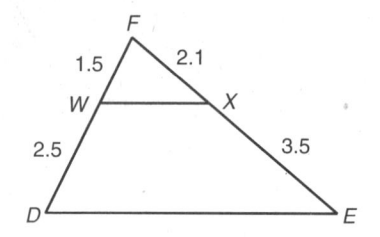

4. $\overline{WX}$ and $\overline{DE}$

$\dfrac{FW}{WD} = \dfrac{1.5}{2.5} = \dfrac{3}{5}$ and $\dfrac{FX}{XE} = \dfrac{2.1}{3.5} = \dfrac{3}{5}$. Because $\dfrac{FW}{WD} = \dfrac{FX}{XE}$, $\overline{WX} \parallel \overline{DE}$ by the Conv. of the △ Proportionality Thm.

Find each length.

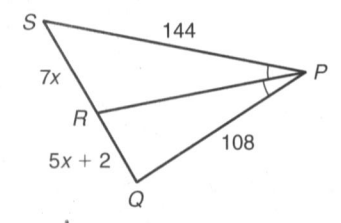

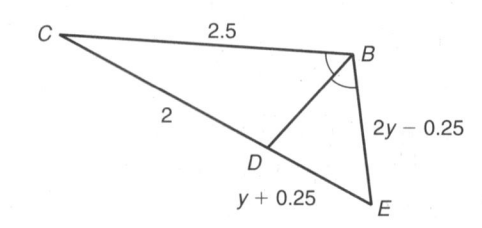

5. SR and RQ ___SR = 56; RQ = 42___

6. BE and DE ___BE = 1.25; DE = 1___

7. In △ABC, $\overline{BD}$ bisects ∠ABC and $\overline{AD} \cong \overline{CD}$. Tell what kind of △ABC must be. ___isosceles___

Name _____ Date _____ Class _____

Practice B
LESSON 7-5 Using Proportional Relationships

Refer to the figure for Exercises 1–3. A city is planning an outdoor concert for an Independence Day celebration. To hold speakers and lights, a crew of technicians sets up a scaffold with two platforms by the stage. The first platform is 8 feet 2 inches off the ground. The second platform is 7 feet 6 inches above the first platform. The shadow of the first platform stretches 6 feet 3 inches across the ground.

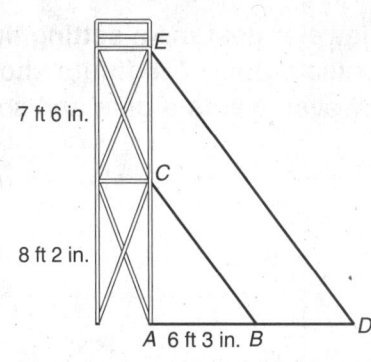

1. Explain why △ABC is similar to △ADE.
   (Hint: The sun's rays are parallel.)

   Possible answer: Because the sun's rays are parallel, $\overline{BC} \parallel \overline{DE}$. ∠ABC and ∠ADE are congruent corresponding angles, and ∠A is common to both triangles. So △ABC ~ △ADE by AA ~.

2. Find the length of the shadow of the second platform in feet and inches to the nearest inch.   **5 ft 9 in.**

3. A 5-foot-8-inch-tall technician is standing on top of the second platform. Find the length of the shadow the scaffold and the technician cast in feet and inches to the nearest inch.   **16 ft 4 in.**

Refer to the figure for Exercises 4–6. Ramona wants to renovate the kitchen in her house. The figure shows a blueprint of the new kitchen drawn to a scale of 1 cm : 2 ft. Use a centimeter ruler and the figure to find each actual measure in feet.

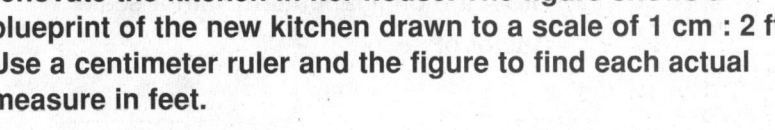

4. width of the kitchen   **10 ft**

5. length of the kitchen   **14 ft**

6. width of the sink   **2 ft**

7. area of the pantry   **12 ft²**

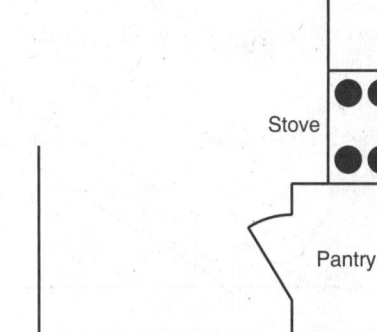

Given that DEFG ~ WXYZ, find each of the following.

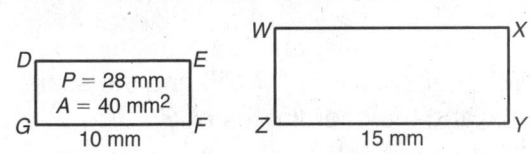

8. perimeter of WXYZ   **42 mm**

9. area of WXYZ   **90 mm²**

Name _____ Date _____ Class _____

LESSON 7-6 Practice B
Dilations and Similarity in the Coordinate Plane

A jeweler designs a setting that can hold a gem in the shape of a parallelogram. The figure shows the outline of the gem. The client, however, wants a gem and setting that is slightly larger.

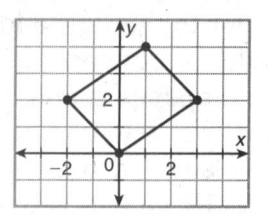

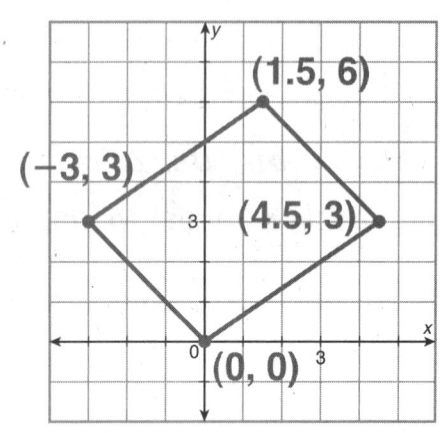

1. Draw the gem after a dilation with a scale factor of $\frac{3}{2}$.

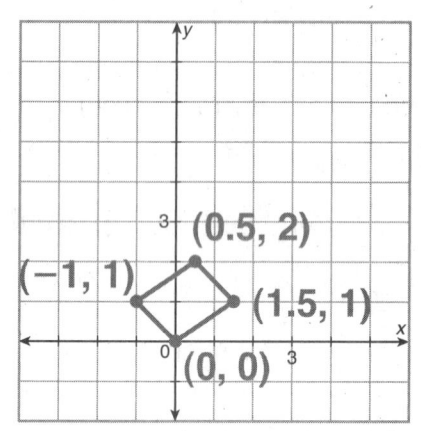

2. The client is so pleased with her ring that she decides to have matching but smaller earrings made using the same pattern. Draw the gem after a dilation from the original pattern with a scale factor of $\frac{1}{2}$.

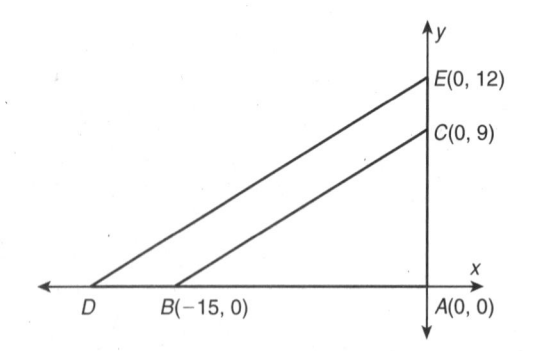

3. Given that $\triangle ABC \sim \triangle ADE$, find the scale factor and the coordinates of D.

   $\frac{4}{3}$; (−20, 0)

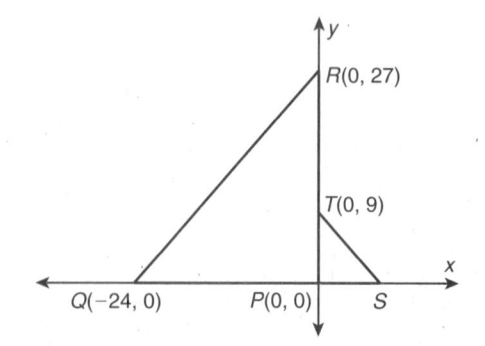

4. Given that $\triangle PQR \sim \triangle PST$, find the scale factor and the coordinates of S.

   $\frac{1}{3}$; (8, 0)

Copyright © by Holt, Rinehart and Winston.
All rights reserved.

Holt Geometry

Name _____ Date _____ Class _____

LESSON 8-1 Practice B
Similarity in Right Triangles

Write a similarity statement comparing the three triangles in each diagram.

1.

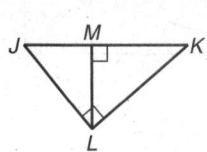

2.

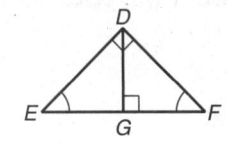

3.

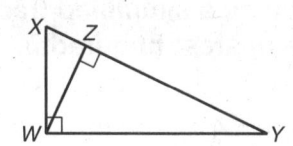

Possible answers:

△JKL ~ △JLM ~ △LKM

△DEF ~ △GED ~ △GDF

△WXY ~ △ZXW ~ △ZWY

Find the geometric mean of each pair of numbers. If necessary, give the answer in simplest radical form.

4. $\frac{1}{4}$ and 4   **1**

5. 3 and 75   **15**

6. 4 and 18   **$6\sqrt{2}$**

7. $\frac{1}{2}$ and 9   **$\frac{3\sqrt{2}}{2}$**

8. 10 and 14   **$2\sqrt{35}$**

9. 4 and 12.25   **7**

Find x, y, and z.

10.

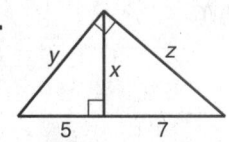

11.

12.

$\sqrt{35}$; $2\sqrt{15}$; $2\sqrt{21}$

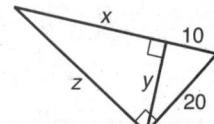

30; $10\sqrt{3}$; $20\sqrt{3}$

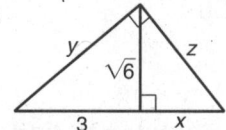

2; $\sqrt{15}$; $\sqrt{10}$

13.

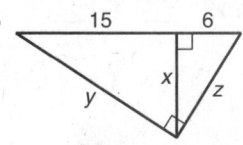

14.

15.

$3\sqrt{10}$; $3\sqrt{35}$; $3\sqrt{14}$

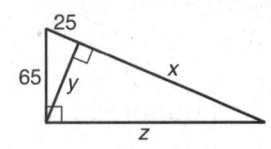

144; 60; 156

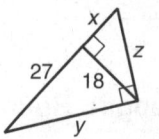

12; $9\sqrt{13}$; $6\sqrt{13}$

16. The Coast Guard has sent a rescue helicopter to retrieve passengers off a disabled ship. The ship has called in its position as 1.7 miles from shore. When the helicopter passes over a buoy that is known to be 1.3 miles from shore, the angle formed by the shore, the helicopter, and the disabled ship is 90°. Determine what the altimeter would read to the nearest foot when the helicopter is directly above the buoy.

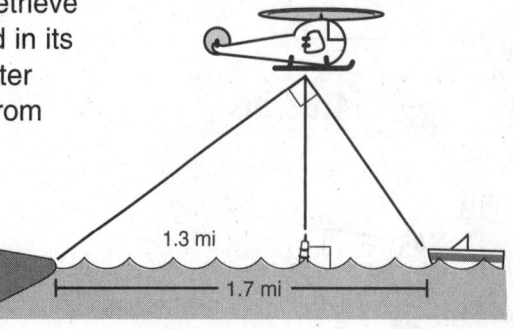

3,807 feet

Use the diagram to complete each equation.

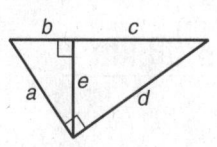

17. $\frac{e}{b} = \frac{\boxed{c}}{e}$

18. $\frac{d}{b+c} = \frac{\boxed{e}}{a}$

19. $\frac{d}{\boxed{c}} = \frac{a}{e}$

Copyright © by Holt, Rinehart and Winston.
All rights reserved.

49

Holt Geometry

Name _____ Date _____ Class _____

LESSON 8-2 Practice B
Trigonometric Ratios

Use the figure for Exercises 1–6. Write each trigonometric ratio as a simplified fraction and as a decimal rounded to the nearest hundredth.

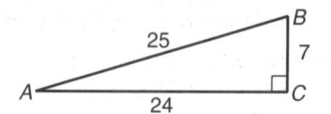

(Triangle: A—C = 24, C—B = 7, A—B = 25, right angle at C)

1. sin A $\dfrac{7}{25}$; 0.28
2. cos B $\dfrac{7}{25}$; 0.28
3. tan B $\dfrac{24}{7}$; 3.43
4. sin B $\dfrac{24}{25}$; 0.96
5. cos A $\dfrac{24}{25}$; 0.96
6. tan A $\dfrac{7}{24}$; 0.29

Use special right triangles to write each trigonometric ratio as a simplified fraction.

7. sin 30° $\dfrac{1}{2}$
8. cos 30° $\dfrac{\sqrt{3}}{2}$
9. tan 45° 1
10. tan 30° $\dfrac{\sqrt{3}}{3}$
11. cos 45° $\dfrac{\sqrt{2}}{2}$
12. tan 60° $\sqrt{3}$

Use a calculator to find each trigonometric ratio. Round to the nearest hundredth.

13. sin 64°   0.90
14. cos 58°   0.53
15. tan 15°   0.27

Find each length. Round to the nearest hundredth.

16. 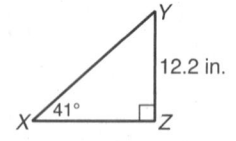 XZ   14.03 in.
17. 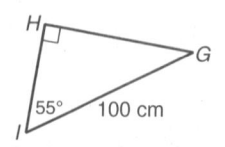 HI   57.36 cm
18. 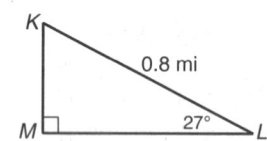 KM   0.36 mi
19. 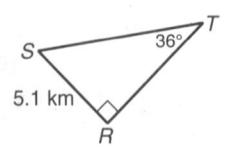 ST   8.68 km
20.  EF   95.41 yd
21. 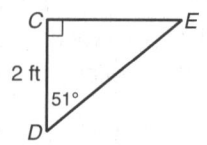 DE   3.18 ft

Copyright © by Holt, Rinehart and Winston.
All rights reserved.

Holt Geometry

Name _____ Date _____ Class _____

LESSON 8-3 Practice B
Solving Right Triangles

Use the given trigonometric ratio to determine which angle of the triangle is ∠A.

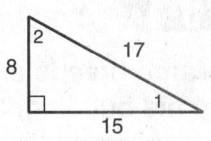

1. $\sin A = \frac{8}{17}$ ∠1
2. $\cos A = \frac{15}{17}$ ∠1
3. $\tan A = \frac{15}{8}$ ∠2
4. $\sin A = \frac{15}{17}$ ∠2
5. $\cos A = \frac{8}{17}$ ∠2
6. $\tan A = \frac{8}{15}$ ∠1

Use a calculator to find each angle measure to the nearest degree.

7. $\sin^{-1}(0.82)$ 55°
8. $\cos^{-1}\left(\frac{11}{12}\right)$ 24°
9. $\tan^{-1}(5.03)$ 79°
10. $\sin^{-1}\left(\frac{3}{8}\right)$ 22°
11. $\cos^{-1}(0.23)$ 77°
12. $\tan^{-1}\left(\frac{1}{9}\right)$ 6°

Find the unknown measures. Round lengths to the nearest hundredth and angle measures to the nearest degree.

13. $AB = 7.74$ in.; $m\angle A = 57°$; $m\angle B = 33°$

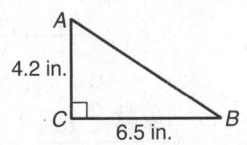

14. $EF = 2.73$ m; $m\angle D = 65°$; $m\angle F = 25°$

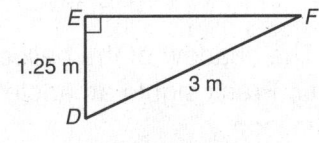

15. $GH = 7.64$ ft; $GI = 7.91$; $m\angle I = 44°$

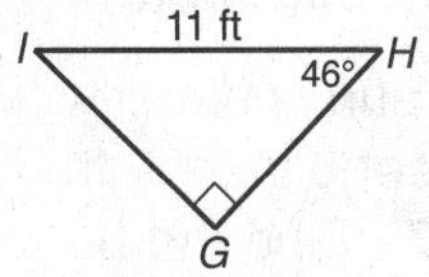

16. $KL = 2.71$ yd; $JK = 2.84$ yd; $m\angle K = 17°$

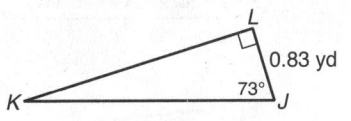

17. $QP = 11.18$ cm; $m\angle Q = 42°$; $m\angle R = 48°$

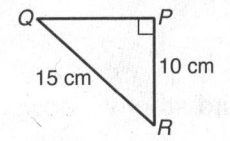

18. $ST = 3.58$ yd; $m\angle S = 12°$; $m\angle T = 78°$

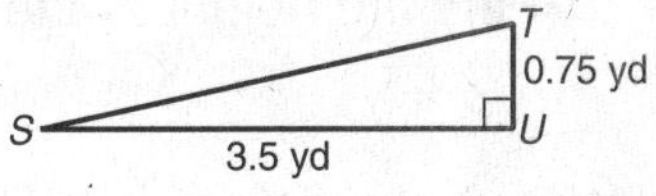

For each triangle, find all three side lengths to the nearest hundredth and all three angle measures to the nearest degree.

19. $B(-2, -4)$, $C(3, 3)$, $D(-2, 3)$

$BC = 8.60$; $BD = 7$; $CD = 5$; $m\angle B = 36°$; $m\angle C = 54°$; $m\angle D = 90°$

20. $L(-1, -6)$, $M(1, -6)$, $N(-1, 1)$

$LM = 2$; $LN = 7$; $MN = 7.28$; $m\angle L = 90°$; $m\angle M = 74°$; $m\angle N = 16°$

21. $X(-4, 5)$, $Y(-3, 5)$, $Z(-3, 4)$

$XY = 1$; $XZ = 1.41$; $YZ = 1$; $m\angle X = 45°$; $m\angle Y = 90°$; $m\angle Z = 45°$

Name _____ Date _____ Class _____

LESSON 8-4 Practice B
Angles of Elevation and Depression

Marco breeds and trains homing pigeons on the roof of his building. Classify each angle as an angle of elevation or an angle of depression.

1. ∠1   __angle of elevation__

2. ∠2   __angle of depression__

3. ∠3   __angle of depression__

4. ∠4   __angle of elevation__

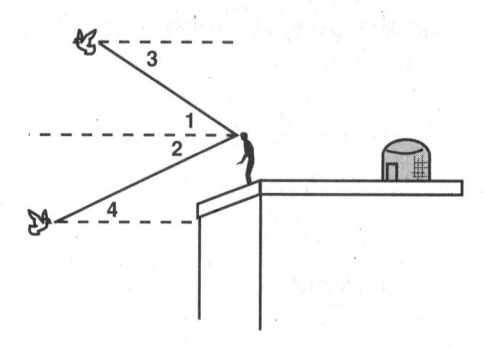

To attract customers to his car dealership, Frank tethers a large red balloon to the ground. In Exercises 5–7, give answers in feet and inches to the nearest inch. (*Note:* Assume the cord that attaches to the balloon makes a straight segment.)

5. The sun is directly overhead. The shadow of the balloon falls 14 feet 6 inches from the tether. Frank sights an angle of elevation of 67°. Find the height of the balloon.   __34 ft 2 in.__

6. Find the length of the cord that tethers the balloon.   __37 ft 1 in.__

7. The wind picks up and the angle of elevation changes to 59°. Find the height of the balloon.   __31 ft 10 in.__

Lindsey shouts down to Pete from her third-story window.

8. Lindsey is 9.2 meters up, and the angle of depression from Lindsey to Pete is 79°. Find the distance from Pete to the base of the building to the nearest tenth of a meter.
   __1.8 m__

9. To see Lindsey better, Pete walks out into the street so he is 4.3 meters from the base of the building. Find the angle of depression from Lindsey to Pete to the nearest degree.
   __65°__

10. Mr. Shea lives in Lindsey's building. While Pete is still out in the street, Mr. Shea leans out his window to tell Lindsey and Pete to stop all the shouting. The angle of elevation from Pete to Mr. Shea is 72°. Tell whether Mr. Shea lives above or below Lindsey.

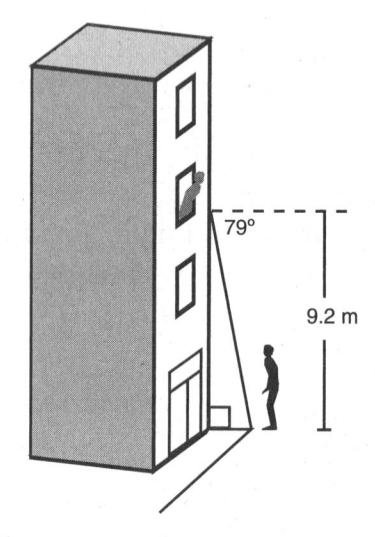

    __Mr. Shea lives above Lindsey.__

Holt Geometry

Name _____ Date _____ Class _____

Practice B
LESSON 8-5 Law of Sines and Law of Cosines

Use a calculator to find each trigonometric ratio. Round to the nearest hundredth.

1. sin 111° __0.93__
2. cos 150° __−0.87__
3. tan 163° __−0.31__

4. sin 92° __1.00__
5. cos 129° __−0.63__
6. tan 99° __−6.31__

7. sin 170° __0.17__
8. cos 96° __−0.10__
9. tan 117° __−1.96__

Use the Law of Sines to find each measure. Round lengths to the nearest tenth and angle measures to the nearest degree.

10. 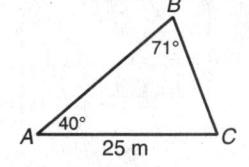 BC __17.0 m__

11. 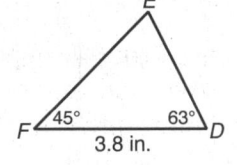 DE __2.8 in.__

12. 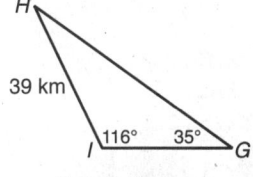 GH __61.1 km__

13. 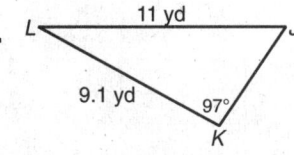 m∠J __55°__

14. 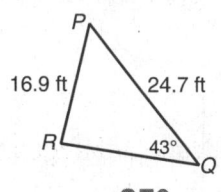 m∠R __85°__

15. 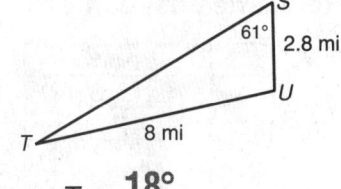 m∠T __18°__

Use the Law of Cosines to find each measure. Round lengths to the nearest tenth and angle measures to the nearest degree.

16. 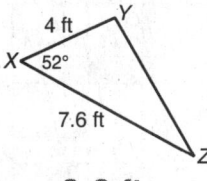 YZ __6.0 ft__

17. 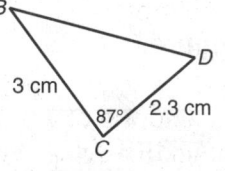 BD __3.7 cm__

18. 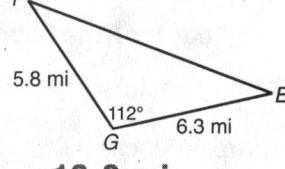 EF __10.0 mi__

19. 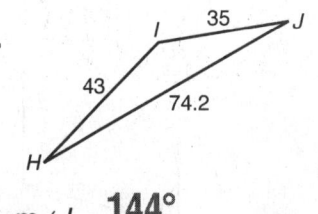 m∠I __144°__

20. 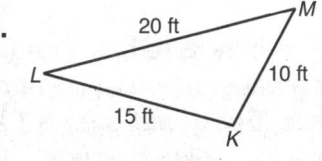 m∠M __47°__

21. 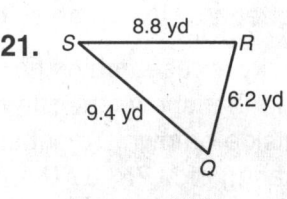 m∠S __40°__

Copyright © by Holt, Rinehart and Winston.
All rights reserved.

Holt Geometry

Name _____ Date _____ Class _____

LESSON 8-6 Practice B
Vectors

Write each vector in component form.

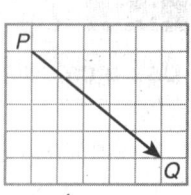

1. $\vec{PQ}$ ⟨5, −4⟩

2. $\vec{EF}$ with $E(-1, 2)$ and $F(-10, -3)$ ⟨−9, −5⟩

3. the vector with initial point $V(7, 3)$ and terminal point $W(0, -1)$ ⟨−7, −4⟩

Draw each vector on a coordinate plane. Find its magnitude to the nearest tenth.

4. ⟨5, 2⟩ 5.4

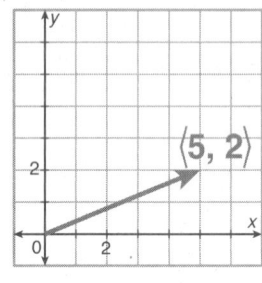

5. ⟨−4, −7⟩ 8.1

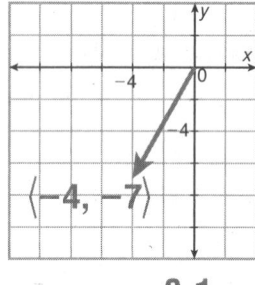

6. ⟨3, −6⟩ 6.7

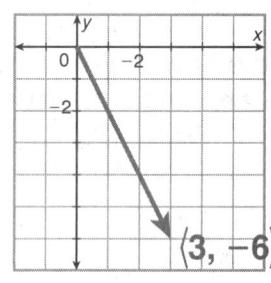

Draw each vector on a coordinate plane. Find the direction of the vector to the nearest degree.

7. ⟨4, 6⟩ 56°

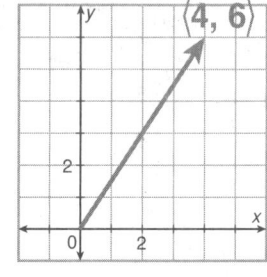

8. ⟨3, 2⟩ 34°

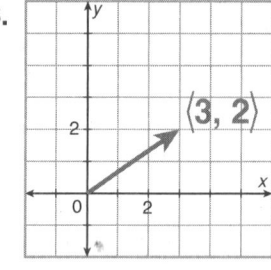

9. ⟨7, 2⟩ 16°

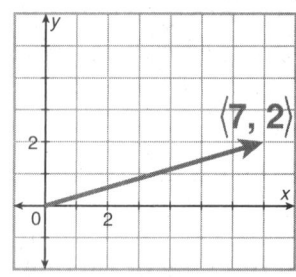

Identify each of the following in the figure.

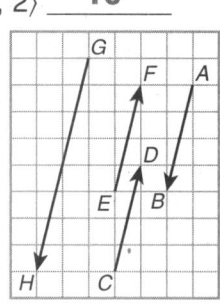

10. equal vectors $\vec{CD}$ and $\vec{EF}$

11. parallel vectors $\vec{AB}$, $\vec{CD}$, $\vec{EF}$, and $\vec{GH}$

In Exercise 12, round directions to the nearest degree and speeds to the nearest tenth.

12. Becky is researching her family history. She has found an old map that shows the site of her great grandparents' farmhouse outside of town. To get to the site, Becky walks for 3.1 km at a bearing of N 75° E. Then she walks 2.2 km due north. Find the distance and direction Becky could have walked to get straight to the site.

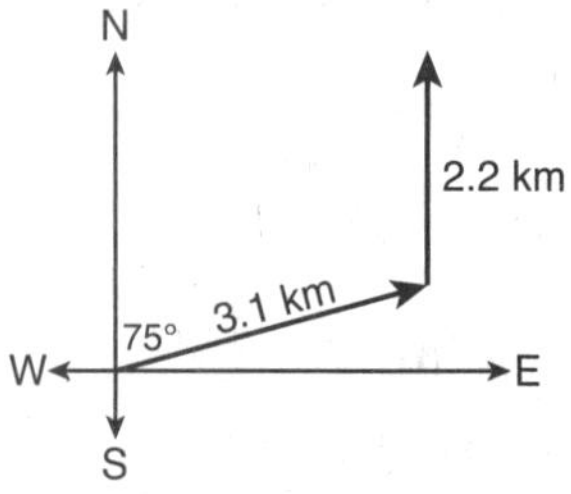

4.2 km; 45° or N 45° E

Copyright © by Holt, Rinehart and Winston.
All rights reserved.

Holt Geometry

Name _____ Date _____ Class _____

LESSON 9-1 Practice B
Developing Formulas for Triangles and Quadrilaterals

Find each measurement.

1. 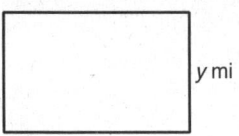 [rectangle, height y mi]

   the perimeter of the rectangle in which $A = 2xy$ mi^2

   $P = (4x + 2y)$ mi

2. 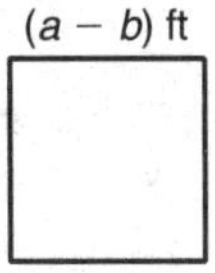 [square, side $(a-b)$ ft]

   the area of the square

   $A = (a-b)^2 = (a^2 - 2ab + b^2)$ ft^2

3. the height of a parallelogram in which $A = 96$ cm^2 and $b = 8x$ cm

   $\dfrac{12}{x}$ cm

4. 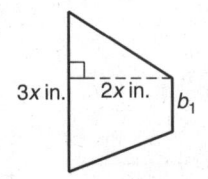 [trapezoid with height 3x in., top segment 2x in., bottom base b_1]

   b_1 of the trapezoid in which $A = 4x^2$ in^2

   $b_1 = x$ in.

5. 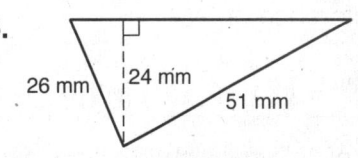 [triangle with sides 26 mm, height 24 mm, base 51 mm]

   the area of the triangle

   $A = 660$ mm^2

6. the area of a trapezoid in which $b_1 = 3a$ km, $b_2 = 6a$ km, and $h = (10 + 4c)$ km

   $A = (45a + 18ac)$ km^2

7. 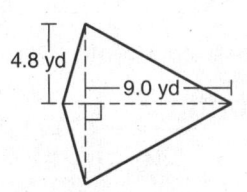 [kite with 4.8 yd and 9.0 yd marked]

   the perimeter of the kite in which $A = 49.92$ yd^2

   $P = 30.4$ yd

8. 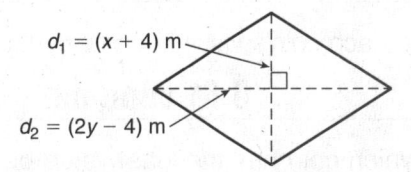 [rhombus with $d_1 = (x+4)$ m and $d_2 = (2y-4)$ m]

   the area of the rhombus

   $A = (xy - 2x + 4y - 8)$ m^2

9. d_2 of the kite in which $d_1 = (a-4)$ ft and $A = (2a^2 - 8a)$ ft^2

   $d_2 = 4a$ ft

Name _____ Date _____ Class _____

LESSON 9-2 Practice B
Developing Formulas for Circles and Regular Polygons

Find each measurement. Give your answers in terms of π.

1. 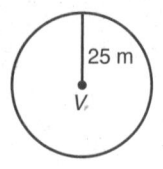 (circle V with radius 25 m)

   the area of ⊙V

   $A = 625\pi$ m²

2. (circle H with diameter 4a in.)

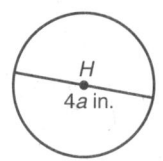

   the area of ⊙H

   $A = 4a^2\pi$ in²

3. 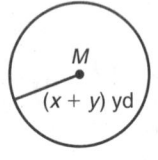 (circle M with radius (x + y) yd)

   the circumference of ⊙M

   $C = (2x + 2y)\pi$ yd

4. 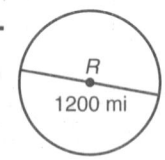 (circle R with diameter 1200 mi)

   the circumference of ⊙R

   $C = 1200\pi$ mi

5. the radius of ⊙D in which $C = 2\pi^2$ cm

   $r = \pi$ cm

6. the diameter of ⊙K in which $A = (x^2 + 2x + 1)\pi$ km²

   $d = (2x + 2)$ km

Stella wants to cover a tabletop with nickels, dimes, or quarters. She decides to find which coin would cost the least to use.

7. Stella measures the diameters of a nickel, a dime, and a quarter. They are 21.2 mm, 17.8 mm, and 24.5 mm. Find the areas of the nickel, the dime, and the quarter. Round to the nearest tenth.

   353.0 mm²; 248.8 mm²; 471.4 mm²

8. Divide each coin's value in cents by the coin's area. Round to the nearest hundredth.

   0.01 cent/mm²; 0.04 cent/mm²; 0.05 cent/mm²

9. Tell which coin has the least value per unit of area.   the nickel

10. Tell about how many nickels would cover a square tabletop that measures 1 square meter. Then find the cost of the coins.

    2833 nickels; $141.65

Find the area of each regular polygon. Round to the nearest tenth.

11. 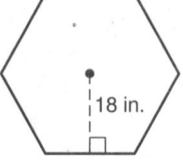 (regular hexagon with apothem 18 in.)

    $A \approx 1122.4$ in²

12. 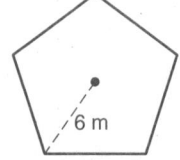 (regular pentagon with radius 6 m)

    $A \approx 85.6$ m²

Copyright © by Holt, Rinehart and Winston.
All rights reserved.

Holt Geometry

Name _____ Date _____ Class _____

LESSON 9-3 Practice B
Composite Figures

Find the shaded area. Round to the nearest tenth if necessary.

1.

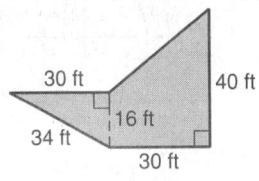

$A = 1080 \text{ ft}^2$

2.

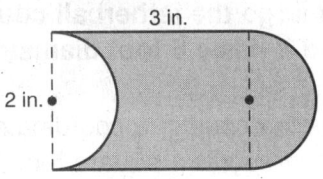

$A = 6 \text{ in}^2$

3.

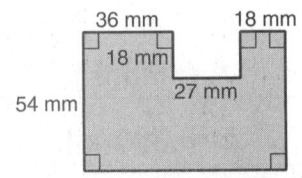

$A = 3888 \text{ mm}^2$

4.

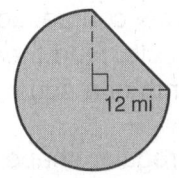

$A \approx 411.3 \text{ mi}^2$

5.

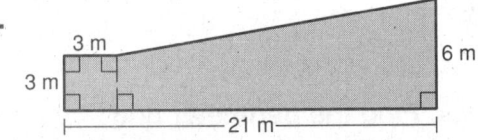

$A = 90 \text{ m}^2$

6.

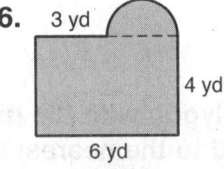

$A \approx 27.5 \text{ yd}^2$

7.

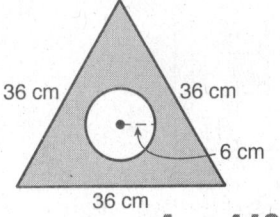

$A \approx 448.1 \text{ cm}^2$

8.

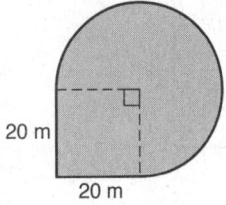

$A \approx 1342.5 \text{ m}^2$

9. Osman broke the unusually shaped picture window in his parents' living room. The figure shows the dimensions of the window. Replacement glass costs $8 per square foot, and there will be a $35 installation fee. Find the cost to replace the window to the nearest cent.

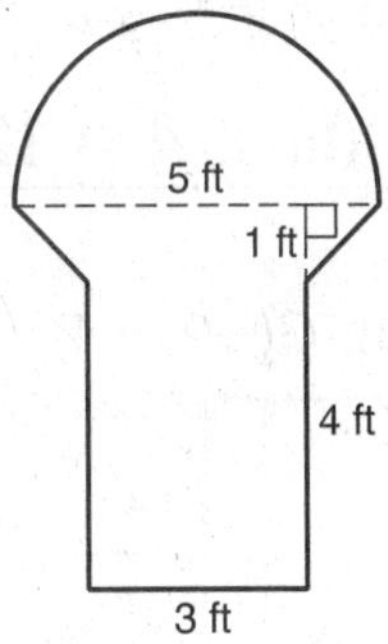

$241.54

Estimate the area of each shaded irregular shape. The grid has squares with side lengths of 1 cm.

10.

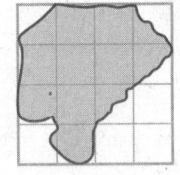

$A \approx 10 \text{ cm}^2$

11.

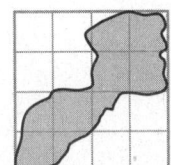

$A \approx 7.5 \text{ cm}^2$

Copyright © by Holt, Rinehart and Winston.
All rights reserved.

Holt Geometry

Name _____ Date _____ Class _____

LESSON 9-4 Practice B
Perimeter and Area in the Coordinate Plane

Lena and her older sister Margie love to play tetherball. They want to find how large the tetherball court is. They measure the court and find it has a 6-foot diameter.

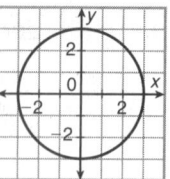

1. Lena sketches the court in a coordinate plane in which each square represents 1 square foot. Estimate the size of the court from the figure.

   Possible answer: $A \approx 30$ ft^2

2. Margie has taken a geometry course, so she knows the formula for the area of a circle. Find the actual area of the court to the nearest tenth of a square foot.

   $A \approx 28.3$ ft^2

3. Estimate the area of the irregular shape.

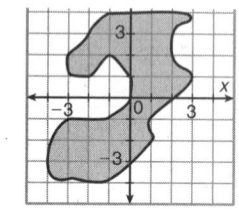

   $A \approx 30$ units^2

Draw and classify each polygon with the given vertices. Find the perimeter and area of the polygon. Round to the nearest tenth if necessary.

4. $A(-2, 3)$, $B(3, 1)$, $C(-2, -1)$, $D(-3, 1)$

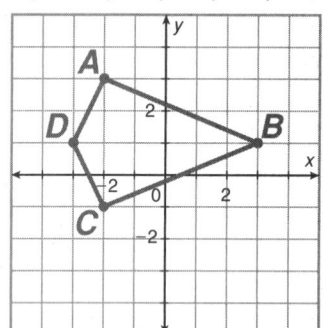

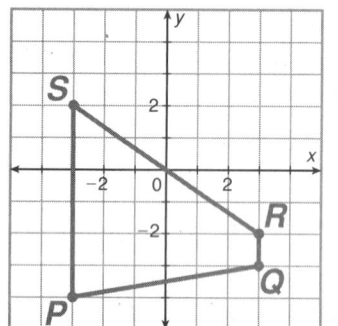

   kite; $P \approx 15.2$ units; $A = 12$ units^2

5. $P(-3, -4)$, $Q(3, -3)$, $R(3, -2)$, $S(-3, 2)$

   trapezoid; $P \approx 20.3$ units; $A = 21$ units^2

6. $E(-4, 1)$, $F(-2, 3)$, $G(-2, -4)$

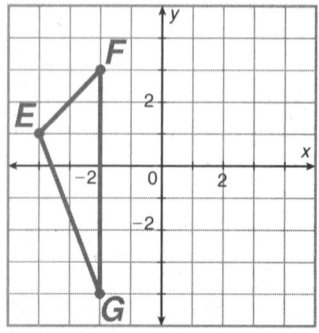

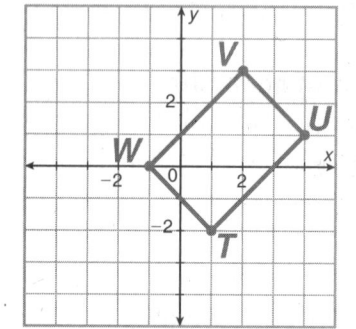

   scalene triangle; $P \approx 15.2$ units; $A = 7$ units^2

7. $T(1, -2)$, $U(4, 1)$, $V(2, 3)$, $W(-1, 0)$

   rectangle; $P \approx 14.1$ units; $A = 12$ units^2

Practice B
9-5 Effects of Changing Dimensions Proportionally

Describe the effect of each change on the area of the given figure.

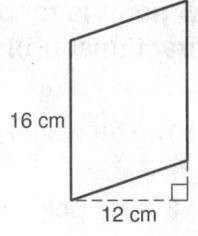

1. The base of the parallelogram is multiplied by $\frac{3}{4}$.
   The area is multiplied by $\frac{3}{4}$.

2. The length of a rectangle with length 12 yd and width 11 yd is divided by 6.
   The area is divided by 6.

3. The base of a triangle with vertices $A(2, 3)$, $B(5, 2)$, and $C(5, 4)$ is doubled.
   The area is doubled.

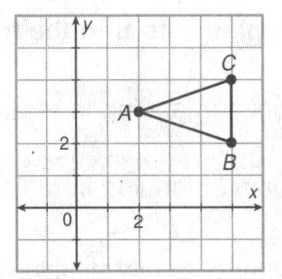

4. The height of a trapezoid with base lengths 4 mm and 7 mm and height 9 mm is multiplied by $\frac{1}{3}$.
   The area is multiplied by $\frac{1}{3}$.

In Exercises 5–8, describe the effect of each change on the perimeter or circumference and the area of the given figure.

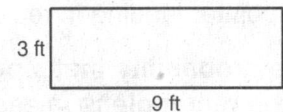

5. The length and width of the rectangle are multiplied by $\frac{4}{3}$.
   The perimeter is multiplied by $\frac{4}{3}$. The area is multiplied by $\frac{16}{9}$.

6. The base and height of a triangle with base 1.5 m and height 6 m are both tripled.
   The perimeter is tripled. The area is multiplied by 9.

7. The radius of a circle with center $(2, 2)$ that passes through $(0, 2)$ is divided by 2.
   The perimeter is divided by 2.
   The area is divided by 4.

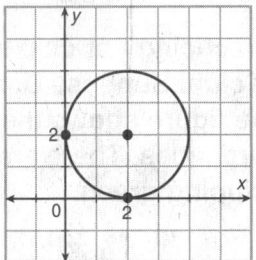

8. The bases and the height of a trapezoid with base lengths 4 in. and 8 in. and height 8 in. are all multiplied by $\frac{1}{8}$.
   The perimeter is multiplied by $\frac{1}{8}$. The area is multiplied by $\frac{1}{64}$.

9. A rhombus has an area of 9 cm^2. The area is multiplied by 5. Describe the effects on the diagonals of the rhombus.
   Both diagonals are multiplied by $\sqrt{5}$.

10. A circle has a circumference of 14π ft. The area is halved. Describe the effects on the circumference of the circle.
    The circumference is divided by $\sqrt{2}$.

Practice B
9-6 Geometric Probability

A point is randomly chosen on $\overline{AD}$. Find the fractional probability of each event.

1. The point is on $\overline{AB}$. __$\frac{5}{12}$__
2. The point is on $\overline{BD}$. __$\frac{7}{12}$__
3. The point is on $\overline{AD}$. __1__
4. The point is not on $\overline{BC}$. __$\frac{2}{3}$__

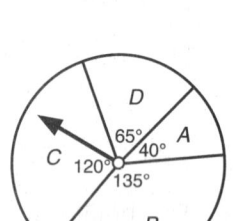

Use the spinner to find the fractional probability of each event.

5. the pointer landing in region C __$\frac{1}{3}$__
6. the pointer landing in region A __$\frac{1}{9}$__
7. the pointer not landing in region D __$\frac{59}{72}$__
8. the pointer landing in regions A or B __$\frac{35}{72}$__

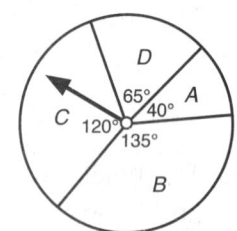

Find the probability that a point chosen randomly inside the rectangle is in each given shape. Round answers to the nearest hundredth.

9. the circle __0.21__
10. the trapezoid __0.20__
11. the circle or the trapezoid __0.41__
12. not the circle and the trapezoid __0.59__

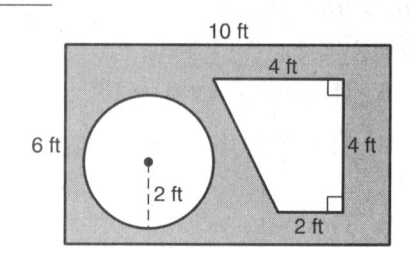

Barb is practicing her chip shots on the chipping green at the local golf club. Suppose Barb's ball drops randomly on the chipping green. The figure shows the chipping green in a grid whose squares have 1-yard sides. There are 18 different 4.5-inch diameter holes on the chipping green.

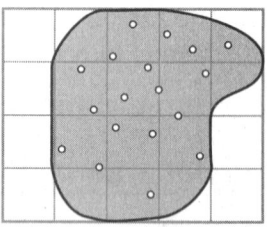

Possible answers based on an 11.5 yd² estimate for the green:

13. Estimate the probability that Barb will chip her ball into any hole. Round to the nearest thousandth. __0.019__

14. Estimate the probability that Barb will chip her ball into the one hole she is aiming for. Round to the nearest thousandth. __0.001__

15. Estimate how many chip shots Barb will have to take to ensure that one goes into a randomly selected hole. __937 shots__

16. Barb is getting frustrated, so her shots are even worse. Now the ball drops randomly anywhere in the grid shown in the figure. Estimate the probability that Barb will miss the chipping green. Round to the nearest thousandth. __0.425__

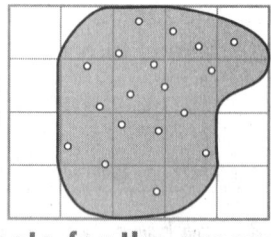

Name _____ Date _____ Class _____

Practice B
LESSON 10-1 Solid Geometry

Classify each figure. Name the vertices, edges, and bases.

1.

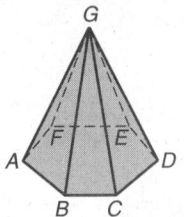

hexagonal pyramid

vertices: A, B, C, D, E, F, and G

edges: $\overline{AB}$, $\overline{BC}$, $\overline{CD}$, $\overline{DE}$, $\overline{EF}$, $\overline{FA}$, $\overline{AG}$, $\overline{BG}$, $\overline{CG}$, $\overline{DG}$, $\overline{EG}$, $\overline{FG}$

base: hexagon ABCDEF

2.

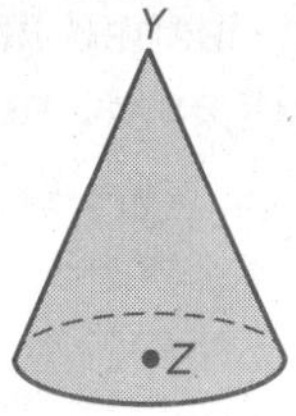

cone

vertices: Y

edges: none

base: ⊙Z

Name the type of solid each object is and sketch an example.

3. a shoe box

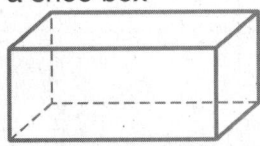

rectangular prism

4. a can of tuna

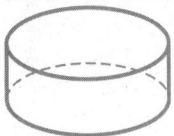

cylinder

Describe the three-dimensional figure that can be made from the given net.

5.

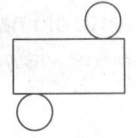

cylinder

6.

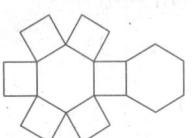

hexagonal prism

7. Two of the nets below make the same solid. Tell which one does not. _____III_____

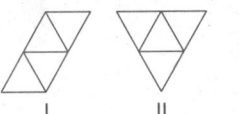

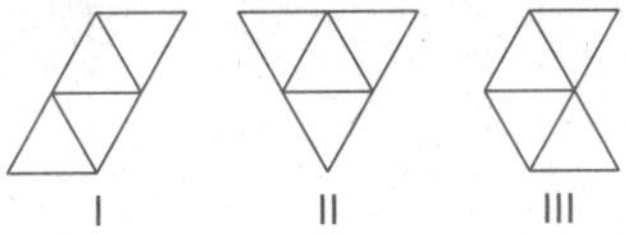

I   II   III

Describe each cross section.

8.

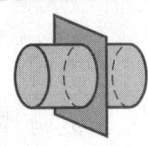

circle

9.

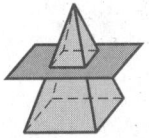

rectangle

10. After completing Exercises 8 and 9, Lloyd makes a conjecture about the shape of any cross section parallel to the base of a solid. Write your own conjecture.

Possible answer: If a cross section intersects a solid parallel to a base, then the cross section has the same shape as the base.

Name _____ Date _____ Class _____

LESSON 10-2 Practice B
Representations of Three-Dimensional Figures

Draw all six orthographic views of each object. Assume there are no hidden cubes. In your answers, use a dashed line to show that the edges touch and a solid line to show that the edges do not touch.

1. 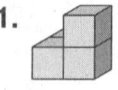 Top Bottom Front

   Back Left Right

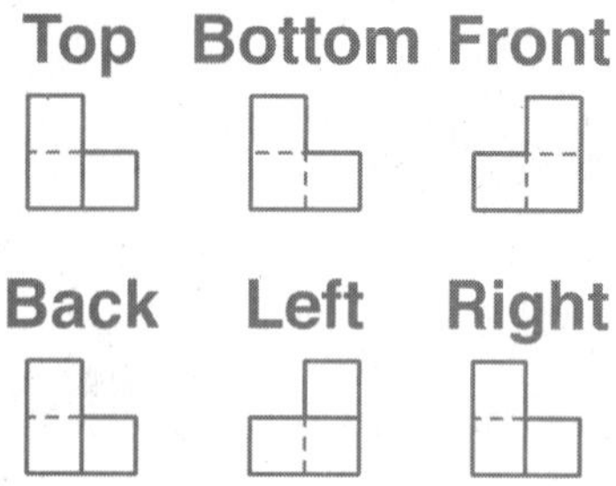

2. 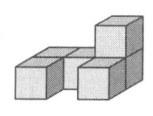 Top Bottom Front

   Back Left Right

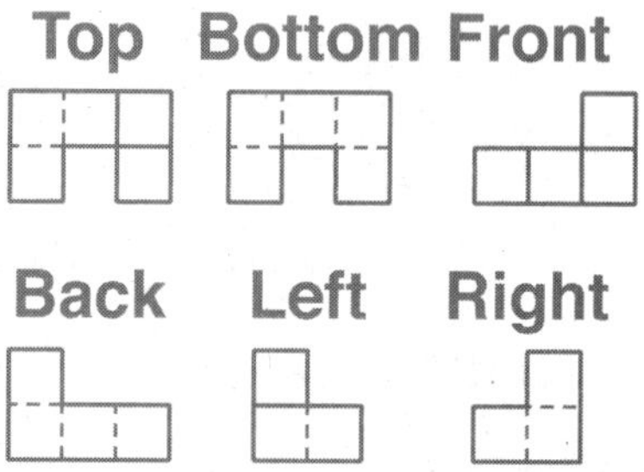

3. Draw an isometric view of the object in Exercise 1.

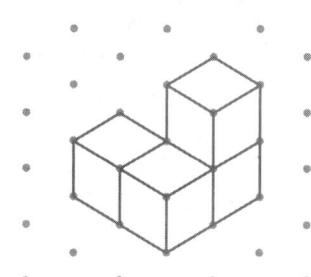

4. Draw an isometric view of the object in Exercise 2.

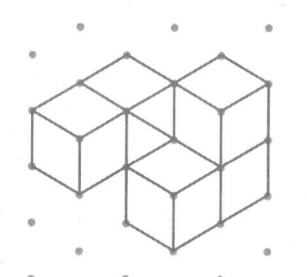

5. Draw a block letter T in one-point perspective.
   Possible answer:

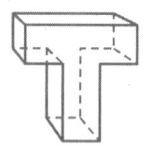

6. Draw a block letter T in two-point perspective. (*Hint:* Draw the vertical line segment that will be closest to the viewer first.) **Possible answer:**

Determine whether each drawing represents the object at right. Assume there are no hidden cubes.

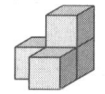

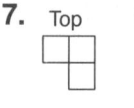

7. 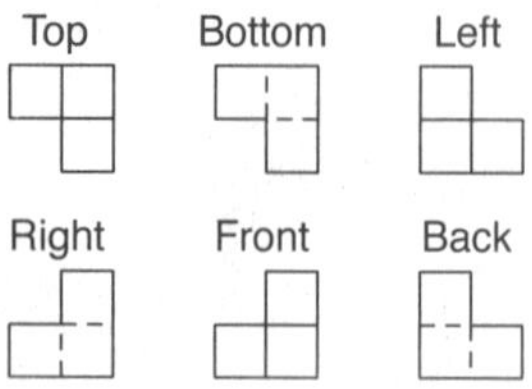 Top Bottom Left

   Right Front Back

   _____yes_____

8.

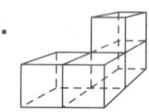

   _____no_____

Copyright © by Holt, Rinehart and Winston.
All rights reserved.

Holt Geometry

Name _____ Date _____ Class _____

LESSON 10-3 Practice B
Formulas in Three Dimensions

Find the number of vertices, edges, and faces of each polyhedron.
Use your results to verify Euler's Formula.

1.

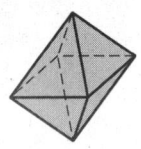

$V = 6$; $E = 12$; $F = 8$;
$6 - 12 + 8 = 2$

2.

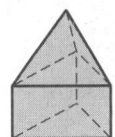

$V = 7$; $E = 12$; $F = 7$;
$7 - 12 + 7 = 2$

Find the unknown dimension in each polyhedron. Round to the nearest tenth.

3. the edge length of a cube with a diagonal of 9 ft — **5.2 ft**

4. the length of a diagonal of a 15-mm-by-20-mm-by-8-mm rectangular prism — **26.2 mm**

5. the length of a rectangular prism with width 2 in., height 18 in., and a 21-in. diagonal — **10.6 in.**

Graph each figure.

6. a square prism with base edge length 4 units, height 2 units, and one vertex at (0, 0, 0)

Possible answer:

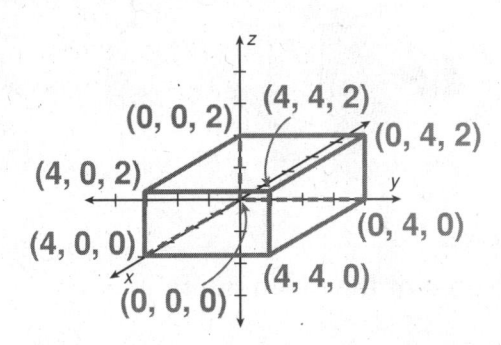

(0, 0, 2), (4, 4, 2), (0, 4, 2), (4, 0, 2), (4, 0, 0), (0, 4, 0), (0, 0, 0), (4, 4, 0)

7. a cone with base diameter 6 units, height 3 units, and base centered at (0, 0, 0)

Possible answer:

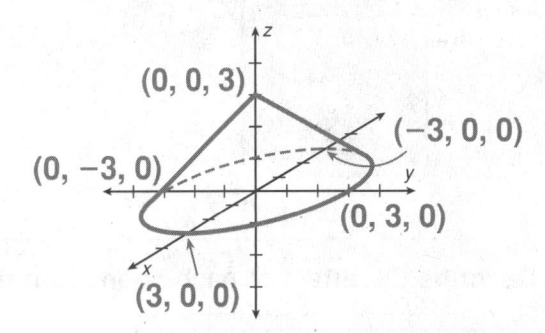

(0, 0, 3), (−3, 0, 0), (0, −3, 0), (0, 3, 0), (3, 0, 0)

Find the distance between the given points. Find the midpoint of the segment with the given endpoints. Round to the nearest tenth if necessary.

8. (1, 10, 3) and (5, 5, 5)

6.7 units; (3, 7.5, 4)

9. (−8, 0, 11) and (2, −6, −17)

30.3 units; (−3, −3, −3)

Name _____ Date _____ Class _____

LESSON 10-4 Practice B
Surface Area of Prisms and Cylinders

Find the lateral area and surface area of each right prism. Round to the nearest tenth if necessary.

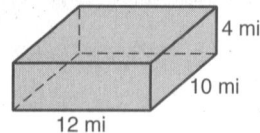

1. [rectangular prism: 4 mi, 10 mi, 12 mi]

   the rectangular prism
   $L = 176$ mi^2; $S = 416$ mi^2

2. [regular pentagonal prism: 7 mm, 2 mm]

   the regular pentagonal prism
   $L = 70$ mm^2; $S = 83.8$ mm^2

3. a cube with edge length 20 inches $L = 1600$ in^2; $S = 2400$ in^2

Find the lateral area and surface area of each right cylinder. Give your answers in terms of π.

[cylinder: 10 cm, 6 cm]

4. $L = 60\pi$ cm^2; $S = 110\pi$ cm^2

5. a cylinder with base area 169π ft^2 and a height twice the radius
   $L = 676\pi$ ft^2; $S = 1014\pi$ ft^2

6. a cylinder with base circumference 8π m and a height one-fourth the radius
   $L = 8\pi$ m^2; $S = 40\pi$ m^2

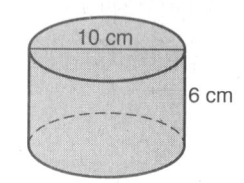

Find the surface area of each composite figure. Round to the nearest tenth.

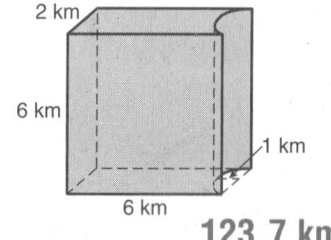

7. [figure: 2 km, 6 km, 1 km, 6 km]

   123.7 km^2

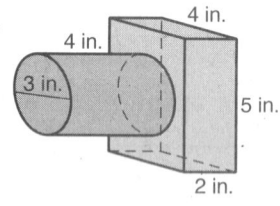

8. [figure: 4 in., 4 in., 3 in., 5 in., 2 in.]

   113.7 in^2

Describe the effect of each change on the surface area of the given figure.

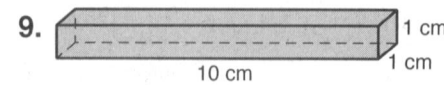

9. [prism: 10 cm, 1 cm, 1 cm]

   The dimensions are multiplied by 12.
   The surface area is multiplied by 144.

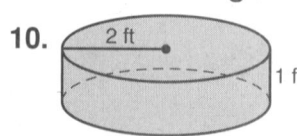

10. [cylinder: 2 ft, 1 ft]

    The dimensions are divided by 4.
    The surface area is divided by 16.

Toby has eight cubes with edge length 1 inch. He can stack the cubes into three different rectangular prisms: 2-by-2-by-2, 8-by-1-by-1, and 2-by-4-by-1. Each prism has a volume of 8 cubic inches.

11. Tell which prism has the smallest surface-area-to-volume ratio.   2-by-2-by-2

12. Tell which prism has the greatest surface-area-to-volume ratio.   8-by-1-by-1

Practice B
10-5 Surface Area of Pyramids and Cones

Find the lateral area and surface area of each regular right solid. Round to the nearest tenth if necessary.

1. 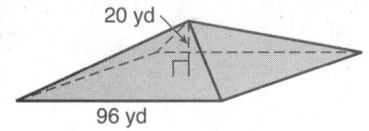 [20 yd, 96 yd]
$L = 9984$ yd^2; $S = 19{,}200$ yd^2

2. 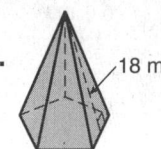 [18 m, 9 m]
$L = 405$ m^2; $S = 544.4$ m^2

3. a regular hexagonal pyramid with base edge length 12 mi and slant height 15 mi
$L = 540$ mi^2; $S = 914.1$ mi^2

Find the lateral area and surface area of each right cone. Give your answers in terms of π.

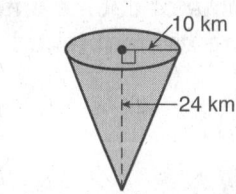

4. [10 km, 24 km]
$L = 260\pi$ km^2; $S = 360\pi$ km^2

5. a right cone with base circumference 14π ft and slant height 3 times the radius
$L = 147\pi$ ft^2; $S = 196\pi$ ft^2

6. a right cone with diameter 240 cm and altitude 35 cm
$L = 15{,}000\pi$ cm^2; $S = 29{,}400\pi$ cm^2

Describe the effect of each change on the surface area of the given figure.

7. 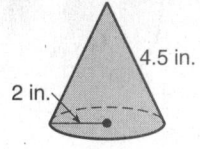 [4.5 in., 2 in.]
The dimensions are multiplied by $\frac{1}{5}$.
The surface area is multiplied by $\frac{1}{25}$.

8. 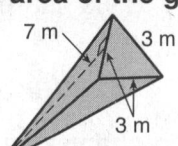 [7 m, 3 m, 3 m]
The dimensions are multiplied by $\frac{3}{2}$.
The surface area is multiplied by $\frac{9}{4}$.

Find the surface area of each composite figure. Round to the nearest tenth if necessary.

9. 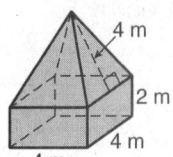 [4 m, 2 m, 4 m, 4 m]
$S = 80$ m^2

10. 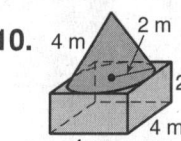 [2 m, 4 m, 2 m, 4 m, 4 m]
$S = 76.6$ m^2

11. The water cooler at Mohammed's office has small conical paper cups for drinking. He uncurls one of the cups and measures the paper. Based on the diagram of the uncurled cup, find the diameter of the cone. [glue tab, 4 in.]
$d = 2$ in.

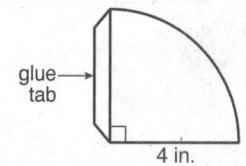

Name _____ Date _____ Class _____

Practice B
LESSON 10-6 Volume of Prisms and Cylinders

Find the volume of each prism. Round to the nearest tenth if necessary.

1. 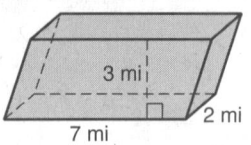 3 mi, 2 mi, 7 mi
the oblique rectangular prism
$V = 42$ mi^3

2. 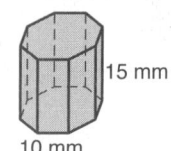 15 mm, 10 mm
the regular octagonal prism
$V \approx 7242.6$ mm^3

3. a cube with edge length 0.75 m $V \approx 0.4$ m^3

Find the volume of each cylinder. Give your answers both in terms of π and rounded to the nearest tenth.

4. 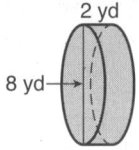 2 yd, 8 yd
$V = 32\pi$ yd^3; $V \approx 100.5$ yd^3

5. 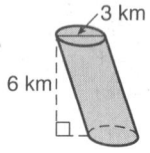 3 km, 6 km
$V = 13.5\pi$ km^3; $V \approx 42.4$ km^3

6. a cylinder with base circumference 18π ft and height 10 ft $V = 810\pi$ ft^3; $V \approx 2544.7$ ft^3

7. CDs have the dimensions shown in the figure. Each CD is 1 mm thick. Find the volume in cubic centimeters of a stack of 25 CDs. Round to the nearest tenth.   0.75 cm, 5.25 cm
$V \approx 278.3$ cm^3

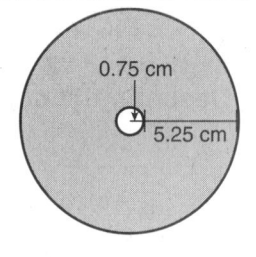

Describe the effect of each change on the volume of the given figure.

8. 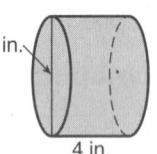 6 in., 4 in.
The dimensions are halved.
The volume is divided by 8.

9. 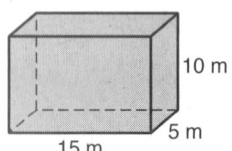 10 m, 5 m, 15 m
The dimensions are divided by 5.
The volume is divided by 125.

Find the volume of each composite figure. Round to the nearest tenth.

10. 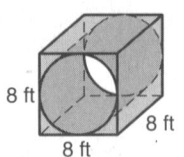 8 ft, 8 ft, 8 ft
$V \approx 109.9$ ft^3

11. 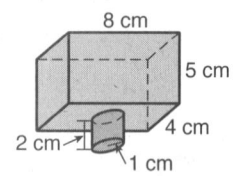 8 cm, 5 cm, 4 cm, 2 cm, 1 cm
$V \approx 166.3$ cm^3

Copyright © by Holt, Rinehart and Winston.
All rights reserved.

Holt Geometry

Practice B
10-7 Volume of Pyramids and Cones

Find the volume of each pyramid. Round to the nearest tenth if necessary.

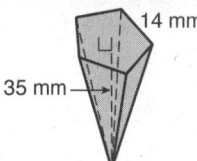

1. 14 mm, 35 mm

   the regular pentagonal pyramid

   $V \approx 3934.2$ mm^3

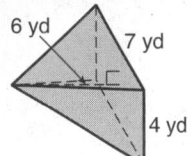

2. 6 yd, 7 yd, 4 yd

   the rectangular right pyramid

   $V = 56$ yd^3

3. Giza in Egypt is the site of the three great Egyptian pyramids. Each pyramid has a square base. The largest pyramid was built for Khufu. When first built, it had base edges of 754 feet and a height of 481 feet. Over the centuries, some of the stone eroded away and some was taken for newer buildings. Khufu's pyramid today has base edges of 745 feet and a height of 471 feet. To the nearest cubic foot, find the difference between the original and current volumes of the pyramid.

   4,013,140 ft^3

Find the volume of each cone. Give your answers both in terms of π and rounded to the nearest tenth.

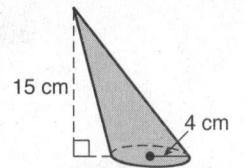

4. 15 cm, 4 cm

   $V = 80\pi$ cm^3; $V \approx 251.3$ cm^3

5. 28 mi, 100 mi

   $V = 25,088\pi$ mi^3; $V \approx 78,816.3$ mi^3

6. a cone with base circumference 6π m and a height equal to half the radius

   $V = 4.5\pi$ m^3; $V \approx 14.1$ m^3

7. Compare the volume of a cone and the volume of a cylinder with equal height and base area.

   The volume of the cone is one-third the volume of the cylinder.

Describe the effect of each change on the volume of the given figure.

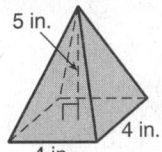

8. 5 in., 4 in., 4 in.

   The dimensions are multiplied by $\frac{2}{3}$.

   The volume is multiplied by $\frac{8}{27}$.

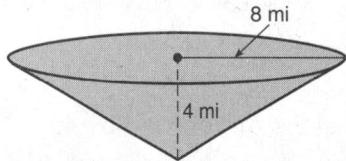

9. 8 mi, 4 mi

   The dimensions are tripled.

   The volume is multiplied by 27.

Find the volume of each composite figure. Round to the nearest tenth.

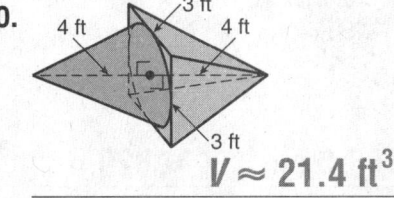

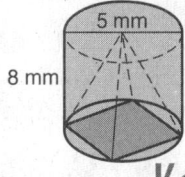

10. 4 ft, 3 ft, 4 ft, 3 ft

    $V \approx 21.4$ ft^3

11. 5 mm, 8 mm

    $V \approx 123.7$ mm^3

Holt Geometry

Name _____ Date _____ Class _____

LESSON 10-8 Practice B
Spheres

Find each measurement. Give your answers in terms of π.

1. [sphere with 18 in. radius]
the volume of the hemisphere
$V = 3888\pi \text{ mm}^3$

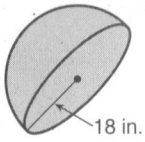

2. [sphere with 26 ft radius]
the volume of the sphere
$V = \dfrac{8788\pi}{3} \text{ ft}^3 = 2929\dfrac{1}{3}\pi \text{ ft}^3$

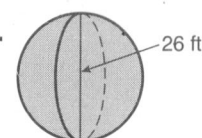

3. the diameter of a sphere with volume $\dfrac{500\pi}{3} \text{ m}^3$
$d = 10 \text{ m}$

4. The figure shows a grapefruit half. The radius to the outside of the rind is 5 cm. The radius to the inside of the rind is 4 cm. The edible part of the grapefruit is divided into 12 equal sections. Find the volume of the half grapefruit and the volume of one edible section. Give your answers in terms of π.
$V = \dfrac{250\pi}{3} \text{ cm}^3; \; V = \dfrac{32\pi}{9} \text{ cm}^3$

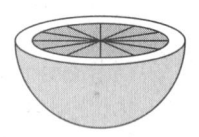

Find each measurement. Give your answers in terms of π.

5. [sphere, A = 121π in²]
the surface area of the sphere
$S = 484\pi \text{ in}^2$

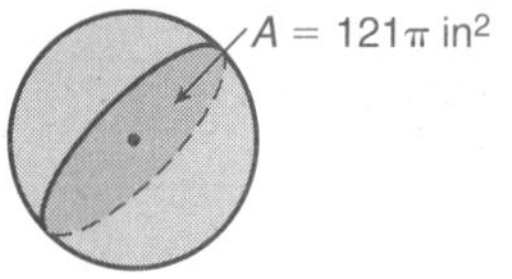

6. [hemisphere, 8 yd]
the surface area of the closed hemisphere and its circular base
$S = 48\pi \text{ yd}^2; \; S = 16\pi \text{ yd}^2$

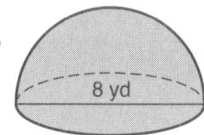

7. the volume of a sphere with surface area $196\pi \text{ km}^2$
$V = \dfrac{1372\pi}{3} \text{ km}^3 = 457\dfrac{1}{3}\pi \text{ km}^3$

Describe the effect of each change on the given measurement of the figure.

8. [sphere, 15 mi]
surface area
The dimensions are divided by 4.
The surface area is divided by 16.

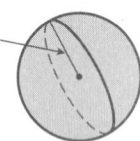

9. [sphere, 36 m]
volume
The dimensions are multiplied by $\dfrac{2}{5}$.
The volume is multiplied by $\dfrac{8}{125}$.

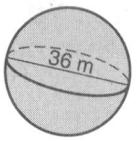

Find the surface area and volume of each composite figure. Round to the nearest tenth.

10. [hemisphere with inscribed cube, 3 in., 3 in., 3 in., 5 in.]
$S \approx 271.6 \text{ in}^2; \; V \approx 234.8 \text{ in}^3$

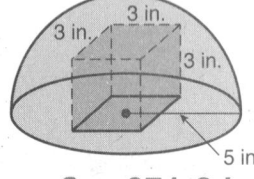

11. [cone with hemisphere, $2\sqrt{34}$ cm, 6 cm]
$S \approx 446.0 \text{ cm}^2; \; V \approx 829.4 \text{ cm}^3$

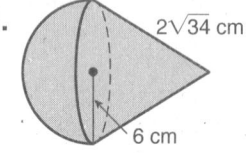

Name _____ Date _____ Class _____

Practice B
LESSON 11-1 Lines That Intersect Circles

Identify each line or segment that intersects each circle.

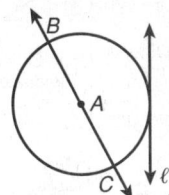

1.

chords: $\overline{BC}$; secant: $\overleftrightarrow{BC}$; tangent: ℓ;
diam.: $\overline{BC}$; radii: $\overline{AB}, \overline{AC}$

2.

chords: $\overline{RQ}, \overline{ST}$; secant: $\overleftrightarrow{ST}$; tangent:
$\overline{UV}$; diam.: $\overline{RQ}$; radii: $\overline{PQ}, \overline{PR}, \overline{PU}$

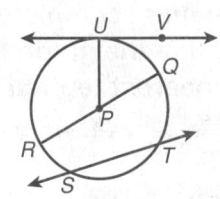

Find the length of each radius. Identify the point of tangency and write the equation of the tangent line at this point.

3.

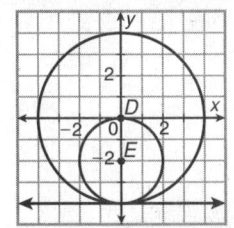

radius of $\odot D$: 4; radius of $\odot E$: 2;
pt. of tangency: $(0, -4)$; eqn. of
tangent line: $y = -4$

4.

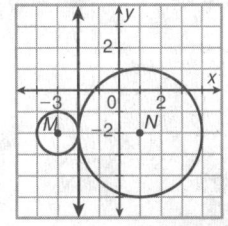

radius of $\odot M$: 1; radius of $\odot N$: 3;
pt. of tangency: $(-2, -2)$; eqn. of
tangent line: $x = -2$

5. The Moon's orbit is not exactly circular, but the average distance from its surface to Earth's surface is 384,000 kilometers. The diameter of the Moon is 3476 kilometers. Find the distance from the surface of Earth to the visible edge of the Moon if the Moon is directly above the observer. Round to the nearest kilometer. (Note: The figure is not drawn to scale.)

385,734 km

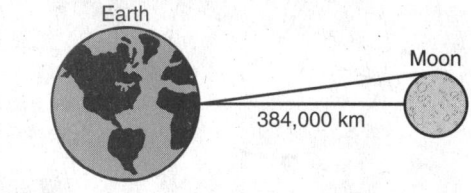

In Exercises 6 and 7, $\overline{EF}$ and $\overline{EG}$ are tangent to $\odot H$. Find EF.

6.

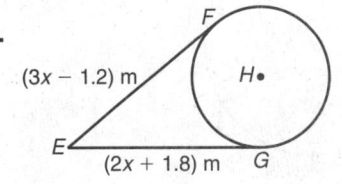

7.8 m

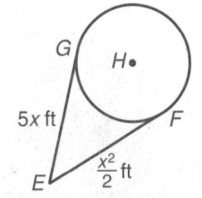

7.

50 ft

Copyright © by Holt, Rinehart and Winston.
All rights reserved.

Holt Geometry

Practice B
11-2 Arcs and Chords

The circle graph shows data collected by the U.S. Census Bureau in 2004 on the highest completed educational level for people 25 and older. Use the graph to find each of the following. Round to the nearest tenth if necessary.

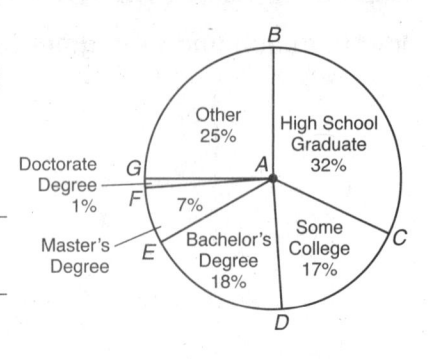

1. m∠CAB __115.2°__
2. m∠DAG __93.6°__
3. m∠EAC __126°__
4. m$\widehat{BG}$ __90°__
5. m$\widehat{GF}$ __3.6°__
6. m$\widehat{BDE}$ __241.2°__

Find each measure.

7. m$\widehat{QS}$ __125°__
   m$\widehat{RQT}$ __227°__

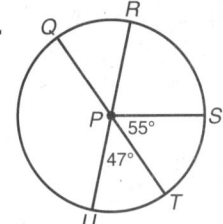

8. m$\widehat{HG}$ __67°__
   m$\widehat{FEH}$ __203°__

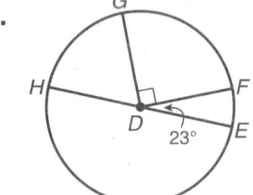

9. Find m∠UTW. __102°__

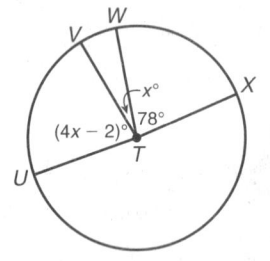

10. ⊙B ≅ ⊙E, and ∠CBD ≅ ∠FEG.
    Find FG. __49 cm__

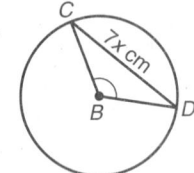

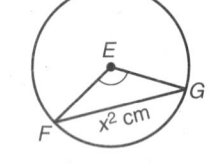

Find each length. Round to the nearest tenth.

11. ZY __76.3 mi__

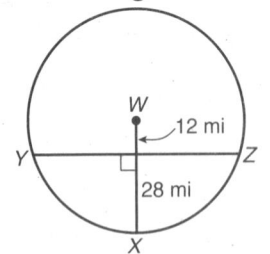

12. EG __4.9 km__

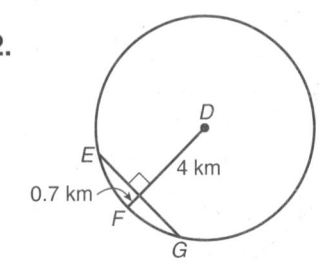

Holt Geometry

Name _____ Date _____ Class _____

Practice B
LESSON 11-3 Sector Area and Arc Length

Find the area of each sector. Give your answer in terms of π and rounded to the nearest hundredth.

1.

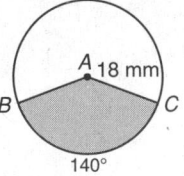

sector BAC 126π mm^2; 395.84 mm^2

2.

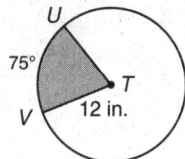

sector UTV 30π in^2; 94.25 in^2

3.

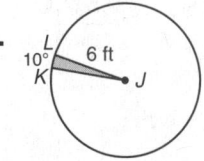

sector KJL π ft^2; 3.14 ft^2

4.

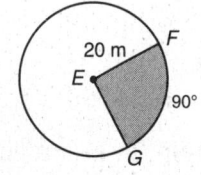

sector FEG 100π m^2; 314.16 m^2

5. The speedometer needle in Ignacio's car is 2 inches long. The needle sweeps out a 130° sector during acceleration from 0 to 60 mi/h. Find the area of this sector. Round to the nearest hundredth.

4.54 in^2

Find the area of each segment to the nearest hundredth.

6.

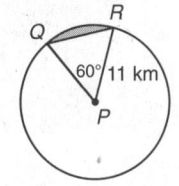

10.96 km^2

7.

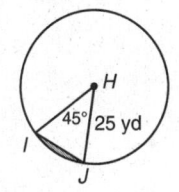

24.47 yd^2

8.

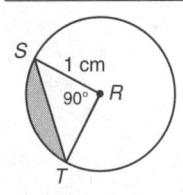

0.29 cm^2

9.

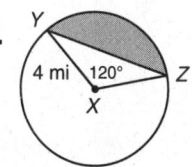

9.83 mi^2

Find each arc length. Give your answer in terms of π and rounded to the nearest hundredth.

10.

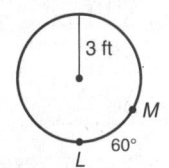

π ft; 3.14 ft

11.

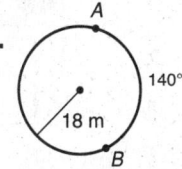

14π m; 43.98 m

12. an arc with measure 45° in a circle with radius 2 mi

$\dfrac{\pi}{2}$ mi; 1.57 mi

13. an arc with measure 120° in a circle with radius 15 mm

10π mm; 31.42 mm

Copyright © by Holt, Rinehart and Winston.
All rights reserved.

Holt Geometry

Name _____ Date _____ Class _____

LESSON 11-4 Practice B
Inscribed Angles

Find each measure.

1. 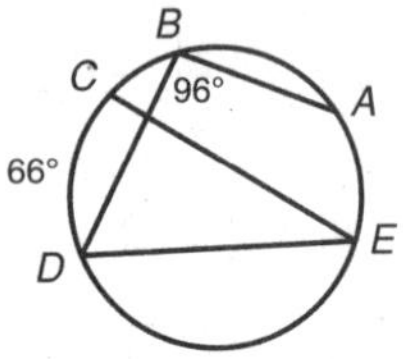 m∠CED = __33°__
   mDEA = __192°__

2. 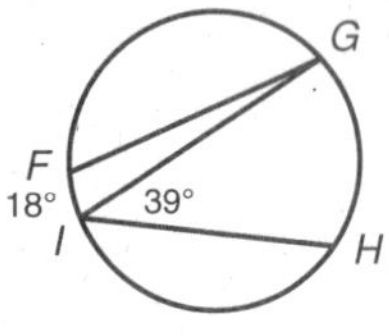 m∠FGI = __9°__
   mGH = __78°__

3. 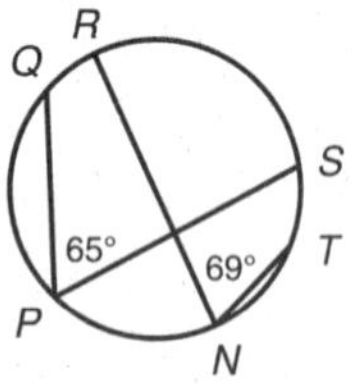 mQRS = __130°__
   mTSR = __138°__

4. 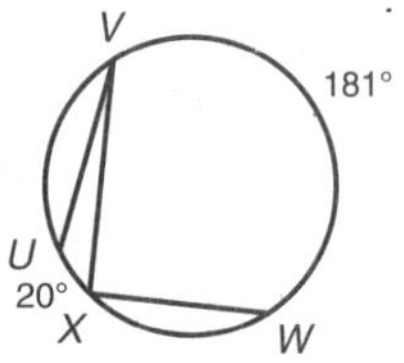 m∠XVU = __10°__
   m∠VXW = __90.5°__

5. A circular radar screen in an air traffic control tower shows these flight paths. Find m∠LNK.   __73°__

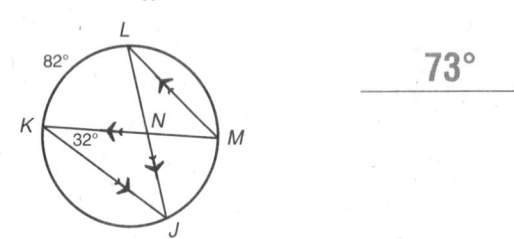

Find each value.

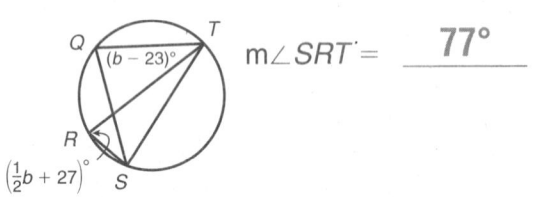

6. 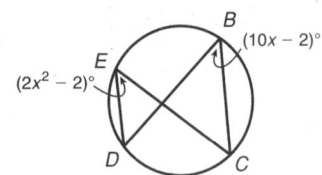 m∠CED = __48°__

7. 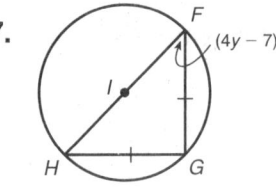 y = __13__

8. 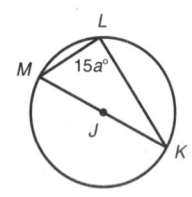 a = __6__

9. m∠SRT = __77°__

Find the angle measures of each inscribed quadrilateral.

10. 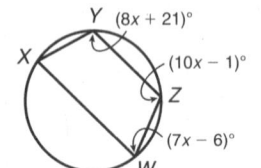 m∠X = __71°__
    m∠Y = __109°__
    m∠Z = __109°__
    m∠W = __71°__

11. 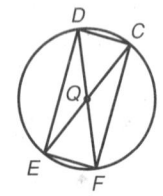 m∠C = __90°__
    m∠D = __90°__
    m∠E = __90°__
    m∠F = __90°__

12. 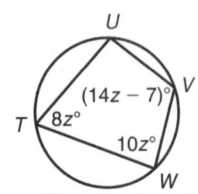 m∠T = __68°__
    m∠U = __95°__
    m∠V = __112°__
    m∠W = __85°__

13. 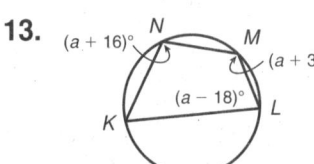 m∠K = __59°__
    m∠L = __73°__
    m∠M = __121°__
    m∠N = __107°__

Name _____ Date _____ Class _____

Practice B
LESSON 11-5 Angle Relationships in Circles

Find each measure.

1. 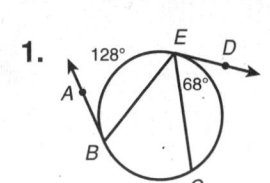 m∠ABE = __64°__
   m$\overarc{BC}$ = __96°__

2. 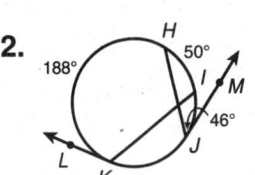 m∠LKI = __119°__
   m$\overarc{IJ}$ = __42°__

3. 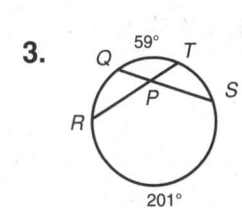 m∠RPS = __130°__

4. 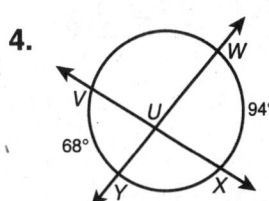 m∠YUX = __99°__

Find the value of x.

5. 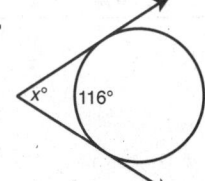 __64__

6. 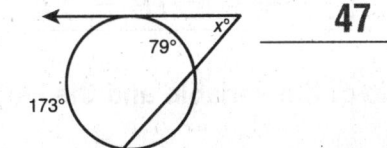 __47__

7. 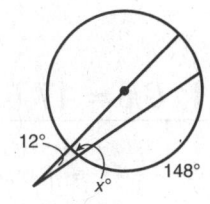 __8__

8. 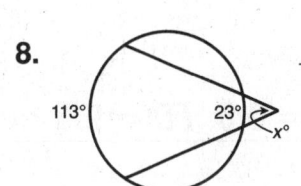 __45__

9. The figure shows a spinning wheel. The large wheel is turned by hand or with a foot trundle. A belt attaches to a small bobbin that turns very quickly. The bobbin twists raw materials into thread, twine, or yarn. Each pair of spokes intercepts a 30° arc. Find the value of x.

__60__

Find each measure.

10. 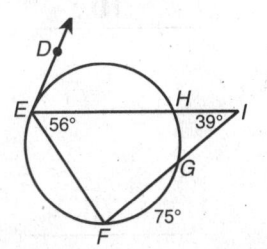 m∠DEI = __66.5°__
    m$\overarc{EF}$ = __115°__

11. 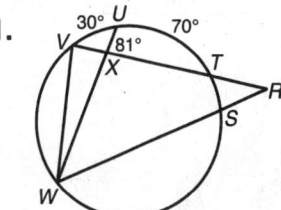 m∠WVR = __84°__
    m$\overarc{TUW}$ = __192°__

Holt Geometry

Name _____ Date _____ Class _____

LESSON 11-6 Practice B
Segment Relationships in Circles

Find the value of the variable and the length of each chord.

1. 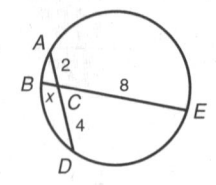 $x = 1;\ AD = 6;\ BE = 9$

2. 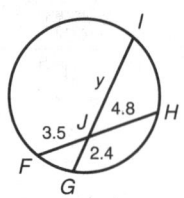 $y = 7;\ FH = 8.3;\ GI = 9.4$

3. 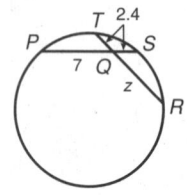 $z = 7;\ PS = 9.4;\ TR = 9.4$

4. 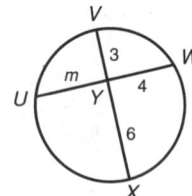 $m = 4.5;\ UW = 8.5;\ VX = 9$

Find the value of the variable and the length of each secant segment.

5. 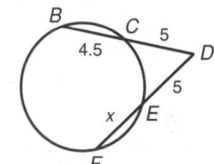 $x = 4.5;\ BD = 9.5;\ FD = 9.5$

6. 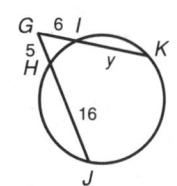 $y = 11.5;\ GJ = 21;\ GK = 17.5$

7. 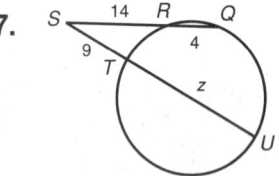 $z = 19;\ SQ = 18;\ SU = 28$

8. 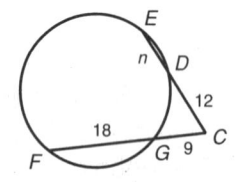 $n = 8.25;\ CE = 20.25;\ CF = 27$

Find the value of the variable. Give answers in simplest radical form if necessary.

9. 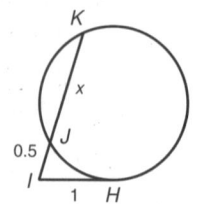 1.5

10. 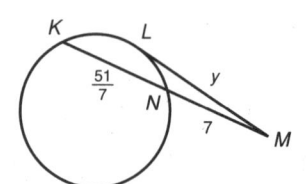 10

11. 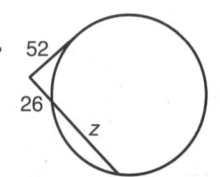 78

12. 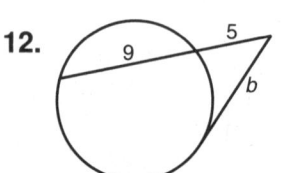 $\sqrt{70}$

Copyright © by Holt, Rinehart and Winston.
All rights reserved.

Holt Geometry

Name _____ Date _____ Class _____

Practice B
LESSON 11-7 Circles in the Coordinate Plane

Write the equation of each circle.

1. ⊙X centered at the origin with radius 10 $x^2 + y^2 = 100$

2. ⊙R with center R(−1, 8) and radius 5 $(x + 1)^2 + (y − 8)^2 = 25$

3. ⊙P with center P(−5, −5) and radius $2\sqrt{5}$ $(x + 5)^2 + (y + 5)^2 = 20$

4. ⊙O centered at the origin that passes through (9, −2) $x^2 + y^2 = 85$

5. ⊙B with center B(0, −2) that passes through (−6, 0) $x^2 + (y + 2)^2 = 40$

6. ⊙F with center F(11, 4) that passes through (−2, 5). $(x − 11)^2 + (y − 4)^2 = 170$

Graph each equation.

7. $x^2 + y^2 = 25$

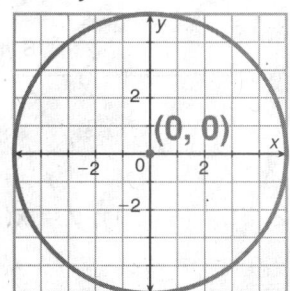

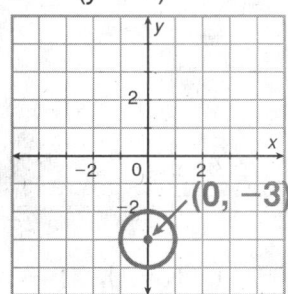

8. $(x + 2)^2 + (y − 1)^2 = 4$

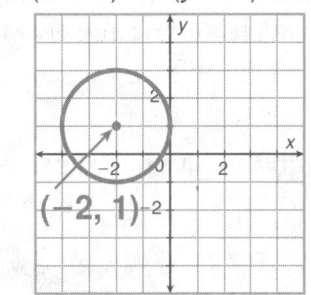

9. $x^2 + (y + 3)^2 = 1$

10. $(x − 1)^2 + (y − 1)^2 = 16$

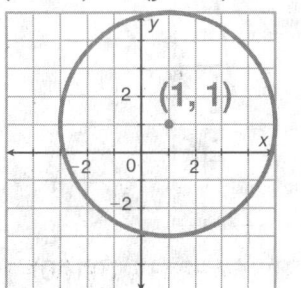

Crater Lake in Oregon is a roughly circular lake. The lake basin formed about 7000 years ago when the top of a volcano exploded in an immense explosion. Hillman Peak, Garfield Peak, and Cloudcap are three mountain peaks on the rim of the lake. The peaks are located in a coordinate plane at H(−4, 1), G(−2, −3), and C(5, −2).

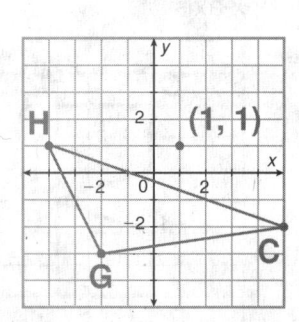

11. Find the coordinates of the center of the lake.

             (1, 1)

12. Each unit of the coordinate plane represents $\frac{3}{5}$ mile. Find the diameter of the lake.

             6 miles

Holt Geometry

Name _____ Date _____ Class _____

LESSON 12-1 Practice B
Reflections

Tell whether each transformation appears to be a reflection.

1. 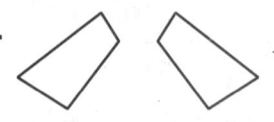 __yes__

2.  __no__

3. 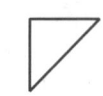  __yes__

4.  __no__

Draw the reflection of each figure across the line.

5.

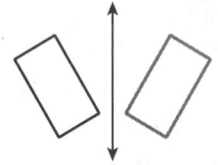

6.

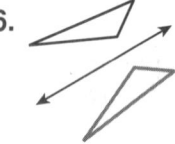

7. Sam is about to dive into a still pool, but some sunlight is reflected off the surface of the water into his eyes. On the figure, plot the exact point on the water's surface where the sunlight is reflected at Sam.

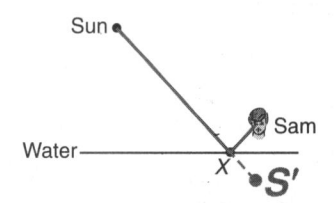

Reflect the figure with the given vertices across the given line.

8. $A(4, 4), B(3, -1), C(1, -2)$; y-axis

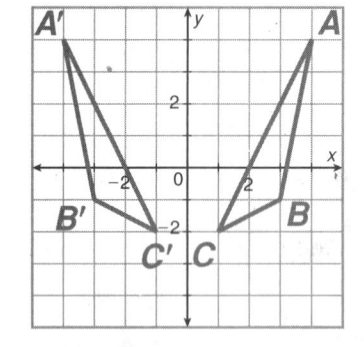

9. $D(-4, -1), E(-2, 3), F(-1, 1)$; $y = x$

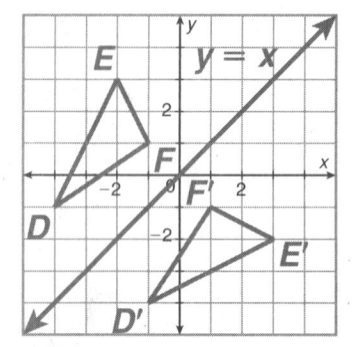

10. $P(1, 3), Q(-2, 3), R(-2, 1), S(1, 0)$; x-axis

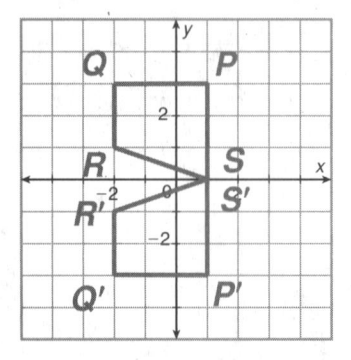

11. $J(3, -4), K(1, -1), L(-1, -1), M(-2, -4)$; $y = x$

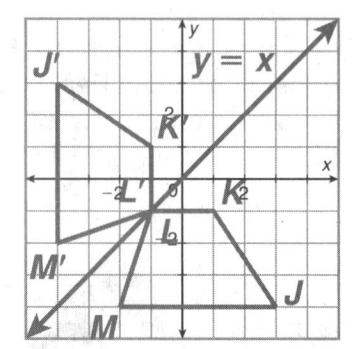

Copyright © by Holt, Rinehart and Winston.
All rights reserved.

Holt Geometry

Name _____ Date _____ Class _____

LESSON 12-2
Practice B
Translations

Tell whether each transformation appears to be a translation.

1. 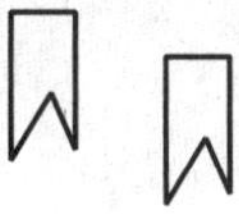 _yes_

2.  _no_

3. _no_

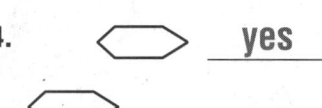

4. _yes_

Draw the translation of each figure along the given vector.

5.

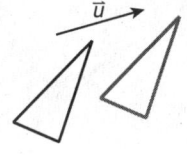

6.

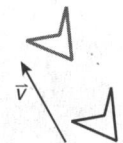

Translate the figure with the given vertices along the given vector.

7. $A(-1, 3), B(1, 1), C(4, 4); \langle 0, -5 \rangle$

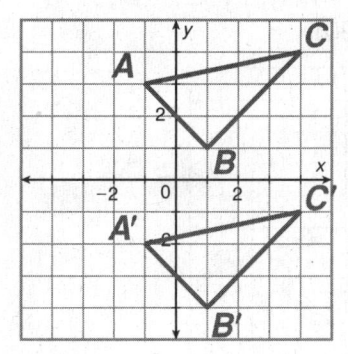

8. $P(-1, 2), Q(0, 3), R(1, 2), S(0, 1); \langle 1, 0 \rangle$

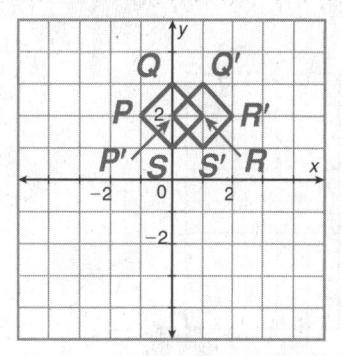

9. $L(3, 2), M(1, -3), N(-2, -2); \langle -2, 3 \rangle$

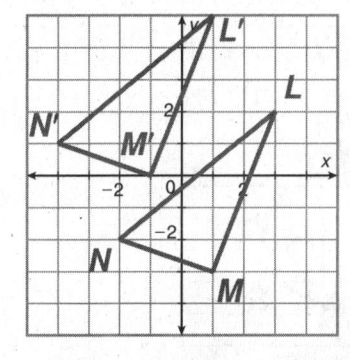

10. $D(2, -2), E(2, -4), F(1, -4), G(-2, -2); \langle 2, 5 \rangle$

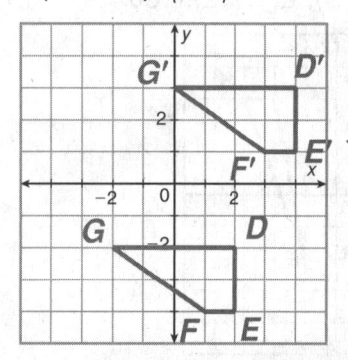

11. A builder is trying to level out some ground with a front-end loader. He picks up some excess dirt at (9, 16) and then maneuvers through the job site along the vectors $\langle -6, 0 \rangle, \langle 2, 5 \rangle,$ and $\langle 8, 10 \rangle$ to get to the spot to unload the dirt. Find the coordinates of the unloading point. Find a single vector from the loading point to the unloading point.

(13, 31); $\langle 4, 15 \rangle$

Holt Geometry

Name _____ Date _____ Class _____

LESSON 12-3 Practice B
Rotations

Tell whether each transformation appears to be a rotation.

1.  _yes_

2. 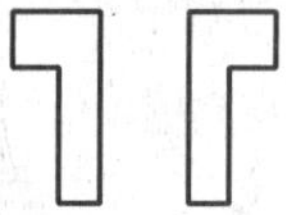 _no_

3. 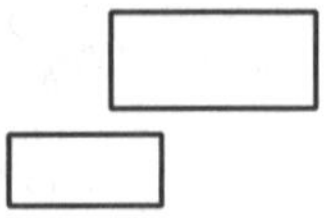 _no_

4. 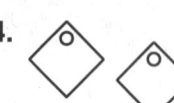 _no_

Draw the rotation of each figure about point P by m∠A.

5.

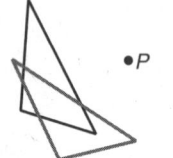

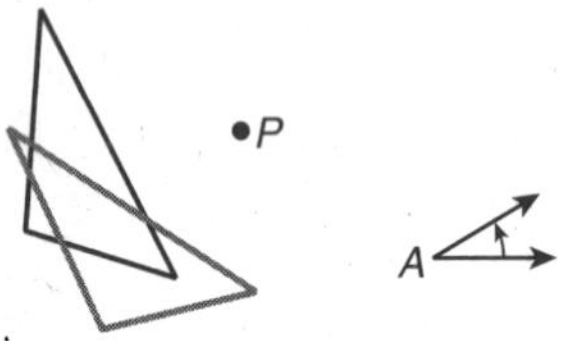

6.

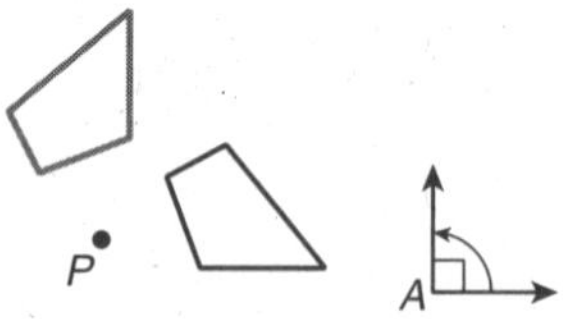

Rotate the figure with the given vertices about the origin using the given angle of rotation.

7. A(−2, 3), B(3, 4), C(0, 1); 90°

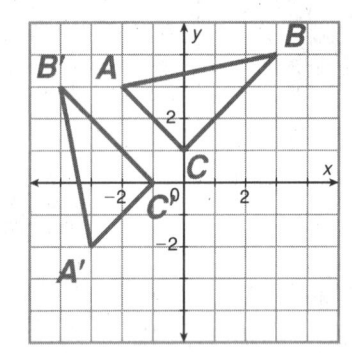

8. D(−3, 2), E(−4, 1), F(−2, −2), G(−1, −1); 90°

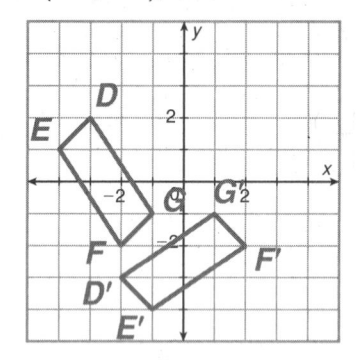

9. J(2, 3), K(3, 3), L(1, −2); 180°

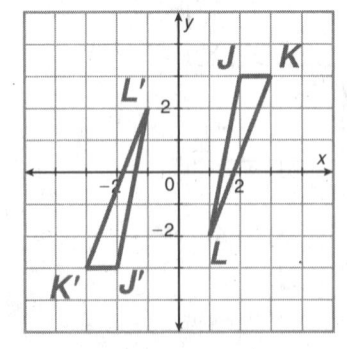

10. P(0, 4), Q(0, 1), R(−2, 2), S(−2, 3); 180°

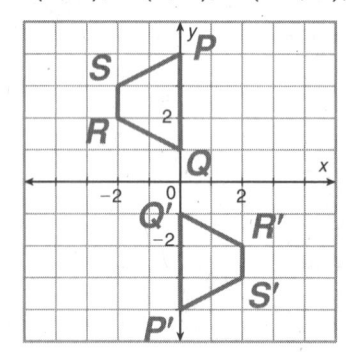

11. The steering wheel on Becky's car has a 15-inch diameter, and its center is at (0, 0). Point X at the top of the wheel has coordinates (0, 7.5). To turn left off her street, Becky must rotate the steering wheel by 300°. Find the coordinates of X when the steering wheel is rotated. Round to the nearest tenth. (Hint: How many degrees short of a full rotation is 300°?) __(6.5, 3.8)__

Name _____ Date _____ Class _____

LESSON 12-4 Practice B
Compositions of Transformations

Draw the result of each composition of isometries.

1. Rotate △XYZ 90° about point P and then translate it along $\vec{v}$.

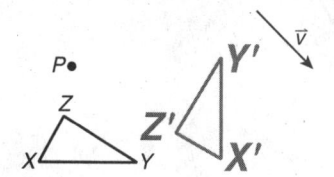

2. Reflect △LMN across line q and then translate it along $\vec{u}$.

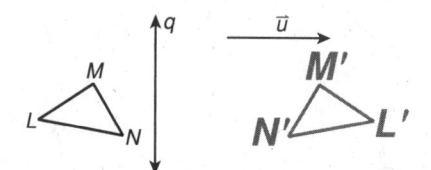

3. ABCD has vertices A(−3, 1), B(−1, 1), C(−1, −1), and D(−3, −1). Rotate ABCD 180° about the origin and then translate it along the vector ⟨1, −3⟩.

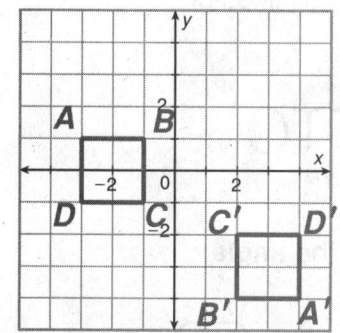

4. △PQR has vertices P(1, −1), Q(4, −1), and R(3, 1). Reflect △PQR across the x-axis and then reflect it across y = x.

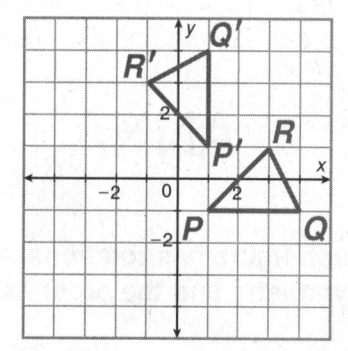

5. Ray draws equilateral △EFG. He draws two lines that make a 60° angle through the triangle's center. Ray wants to reflect △EFG across ℓ_1 and then across ℓ_2. Describe what will be the same and what will be different about the image of △E″F″G″ compared to △EFG.

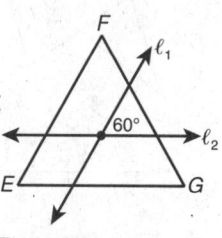

The sides of the image will lie on the sides of the preimage, but the position of the vertices will be different. E″ will coincide with F, F″ with G, and G″ with E.

Draw two lines of reflection that produce an equivalent transformation for each figure.

6. translation: STUV → S'T'U'V'

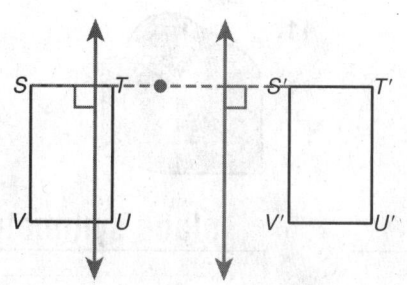

7. rotation with center P: STUV → S'T'U'V'

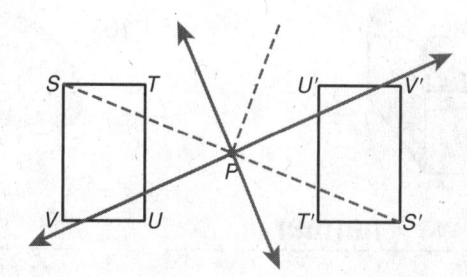

Copyright © by Holt, Rinehart and Winston.
All rights reserved.

Holt Geometry

Name _____ Date _____ Class _____

LESSON 12-5 Practice B
Symmetry

Tell whether each figure has line symmetry. If so, draw all lines of symmetry.

1. 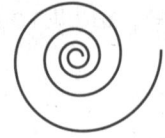 _no_

2. 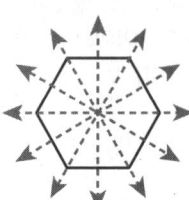 _yes_

3. 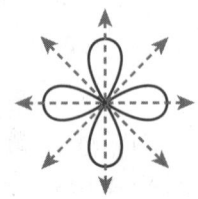 _yes_

4. Anna, Bob, and Otto write their names in capital letters. Draw all lines of symmetry for each whole name if possible.

ANNA   ←B−O−B→   OT|TO

Tell whether each figure has rotational symmetry. If so, give the angle of rotational symmetry and the order of the symmetry.

5.  _yes; 180°; 2_

6. 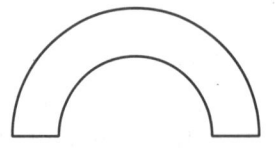 _no_

7.  _yes; 45°; 8_

8. This figure shows the Roman symbol for Earth. Draw all lines of symmetry. Give the angle and order of any rotational symmetry.
   _90°; 4_

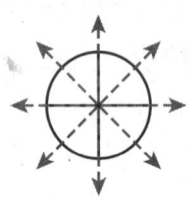

Tell whether each figure has plane symmetry, symmetry about an axis, both, or neither.

9. 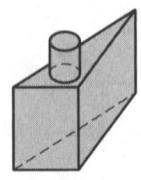 _neither_

10. 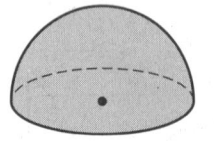 _both_

11. 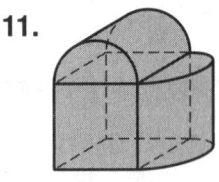 _plane symmetry_

Copyright © by Holt, Rinehart and Winston.
All rights reserved.

Holt Geometry

Name _____ Date _____ Class _____

Practice B
LESSON 12-6 Tessellations

Tell whether each pattern has translation symmetry, glide reflection symmetry, or both.

1.  translation symmetry

2.  both

3. 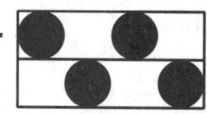 glide reflection symmetry

Use the given figure to create a tessellation. Possible answers:

4.

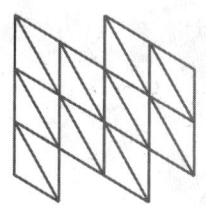

5.

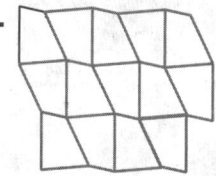

Classify each tessellation as regular, semiregular, or neither.

6.  regular

7. 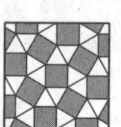 semiregular

8. 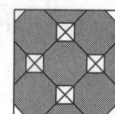 neither

Determine whether the given regular polygon(s) can be used to form a tessellation. If so, draw the tessellation.

9. 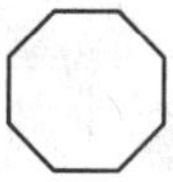 no

10. 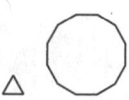 yes

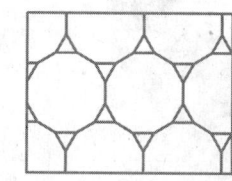

11. 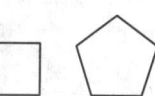 no

Copyright © by Holt, Rinehart and Winston.
All rights reserved.

Holt Geometry

Name _____ Date _____ Class _____

Practice B
LESSON 12-7 Dilations

Tell whether each transformation appears to be a dilation.

1. 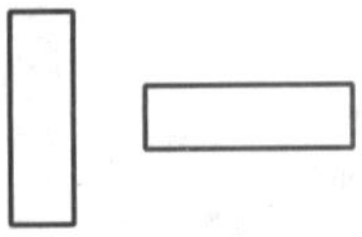 _no_

2. 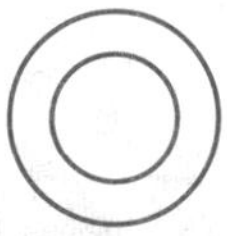 _yes_

3. 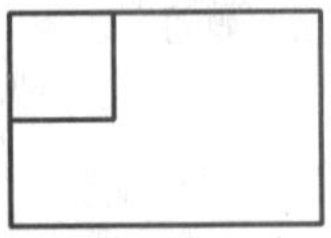 _no_

4. 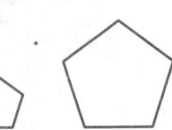 _yes_

Draw the dilation of each figure under the given scale factor with center of dilation P.

5. scale factor: $\frac{1}{2}$

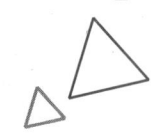

6. scale factor: -2

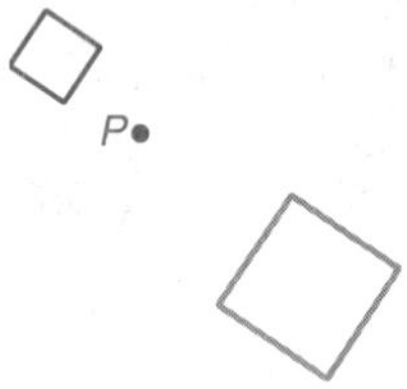

7. A sign painter creates a rectangular sign for Mom's Diner on his computer desktop. The desktop version is 12 inches by 4 inches. The actual sign will be 15 feet by 5 feet. If the capital M in "Mom's" will be 4 feet tall, find the height of the M on his desktop version. $3\frac{1}{5}$ inches

Draw the image of the figure with the given vertices under a dilation with the given scale factor centered at the origin.

8. $A(2, -2)$, $B(2, 3)$, $C(-3, 3)$, $D(-3, -2)$; scale factor: $\frac{1}{2}$

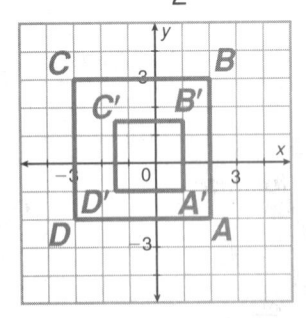

9. $P(-4, 4)$, $Q(-3, 1)$, $R(2, 3)$; scale factor: -1

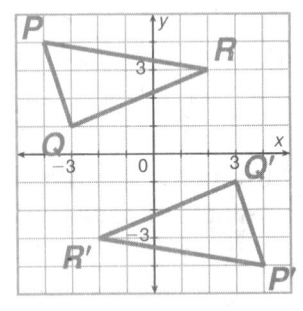

10. $J(0, 2)$, $K(-2, 1)$, $L(0, -2)$, $M(2, -1)$; scale factor: 2

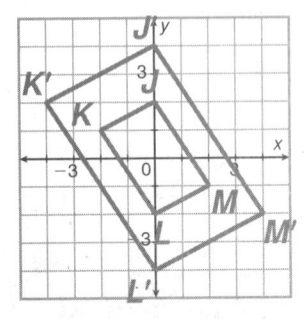

11. $D(0, 0)$, $E(-1, 0)$, $F(-1, -1)$; scale factor: -2

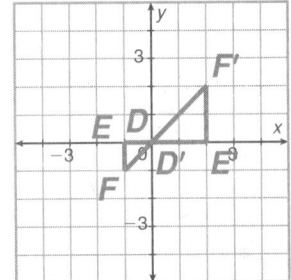

Holt Geometry